Decision and Decision-maker in an Industrial Environment

Decision and Decision-maker in an Industrial Environment

Lamia Berrah
Vincent Clivillé

WILEY

First published 2023 in Great Britain and the United States by ISTE Ltd and John Wiley & Sons, Inc.

Apart from any fair dealing for the purposes of research or private study, or criticism or review, as permitted under the Copyright, Designs and Patents Act 1988, this publication may only be reproduced, stored or transmitted, in any form or by any means, with the prior permission in writing of the publishers, or in the case of reprographic reproduction in accordance with the terms and licenses issued by the CLA. Enquiries concerning reproduction outside these terms should be sent to the publishers at the undermentioned address:

ISTE Ltd
27-37 St George's Road
London SW19 4EU
UK

www.iste.co.uk

John Wiley & Sons, Inc.
111 River Street
Hoboken, NJ 07030
USA

www.wiley.com

© ISTE Ltd 2023

The rights of Lamia Berrah and Vincent Clivillé to be identified as the authors of this work have been asserted by them in accordance with the Copyright, Designs and Patents Act 1988.

Any opinions, findings, and conclusions or recommendations expressed in this material are those of the author(s), contributor(s) or editor(s) and do not necessarily reflect the views of ISTE Group.

Library of Congress Control Number: 2022948736

British Library Cataloguing-in-Publication Data
A CIP record for this book is available from the British Library
ISBN 978-1-78630-730-9

	4.3.3. The AHP method	152
	4.3.4. The MACBETH method	168
4.4. Conclusion		187

Conclusion 189

References 193

Index 207

Contents

Foreword ix

Preface xiii

Chapter 1. Decision and Decision Context 1

   1.1. Introduction. 1
   1.2. A fleeting look at some of the great civilizations. 5
      1.2.1. Method of investigation 5
      1.2.2. The ancient Near-East 6
      1.2.3. Ancient Egypt 10
      1.2.4. India. 12
      1.2.5. China 16
      1.2.6. Ancient Greece 21
      1.2.7. The Arab–Muslim world 25
      1.2.8. The Western world 28
   1.3. Conclusion 33

Chapter 2. Decisions: The Process 37

   2.1. Introduction. 37
   2.2. Why a decision process? 40
   2.3. The notion of process 43
   2.4. Decision-making: rationality or intuition 44
   2.5. The classical theory of rationality. 46
      2.5.1. The search for a numerical value. 46
      2.5.2. Fundamentals 47
      2.5.3. Operational research 48
      2.5.4. Game theory 52

  2.5.5. Taking account of uncertainty: towards procedural rationality . . . 53
  2.5.6. A return to the decision process 56
  2.5.7. Summary and analysis 60
 2.6. Procedural rationality theory 62
  2.6.1. The search for a satisfactory solution 62
  2.6.2. The basics 64
  2.6.3. The school of Herbert Simon 66
  2.6.4. Extensions to Simon's process 74
  2.6.5. Procedural rationality and artificial intelligence 83
  2.6.6. A return to the decision process 84
  2.6.7. Summary and analysis 87
 2.7. Conclusion 89

Chapter 3. The Decision: The Multi-criteria Universe 91

 3.1. Introduction 91
 3.2. Comparing to be able to choose 93
  3.2.1. Intuitive vision 93
  3.2.2. In a single criterion universe 94
  3.2.3. In a multi-criteria universe 94
 3.3. The construction of MCDA 95
 3.4. Vocabulary 98
  3.4.1. Solutions 98
  3.4.2. The criteria 101
  3.4.3. Decision types 107
 3.5. Ordering for comparison 111
  3.5.1. Intuitive vision 111
  3.5.2. The notion of preferences 112
  3.5.3. Preferences and order relationships 114
 3.6. The particular case of Pareto dominance 121
 3.7. Summary 124
 3.8. Conclusion 126

Chapter 4. The Decision: Methods 127

 4.1. Introduction 127
 4.2. Outranking 129
  4.2.1. Principles 129
  4.2.2. Condorcet's method 131
  4.2.3. The ELECTRE III method 134
 4.3. Aggregation 147
  4.3.1. Principles 147
  4.3.2. The Borda count method 150

Foreword

The contribution to the discussion on decisions and decision support that Lamia Berrah and Vincent Clivillé propose in this book is certainly very stimulating and, as we will try to explain, has a particular cultural value in the context of scientific decision support. To be fully appreciated, the content of this book should be placed in the current context of the widespread diffusion of information technologies in our daily life, as well as in the context of reflections on decisions, especially those made over 40 years by researchers in the scientific community of the "European school" of decision-aiding, to which Lamia Berrah and Vincent Clivillé refer explicitly.

We are living in an era in which there is an excess of data and information and growing computing capacity. This large amount of data and considerable computing power are increasingly used in our daily life, for example when we consult the Internet and when we are offered products of all kinds meant to maximize our satisfaction on the basis of preferences we have expressed, even implicitly. The temptation to think of each decision in terms of optimization is ever greater. After all, deciding would simply amount to finding the best solution given our preferences. In reality, this is a deceptive caricature of the decisions that we make in our daily life. We know that the more important the decisions we have to make, the more our hesitations, doubts, uncertainties and questions increase. These hesitations, doubts, uncertainties and questions are present, indeed amplified, even in the case of decisions involving complex structures such as businesses or public bodies, states or supranational organizations. If we then consider society's complex problems, such as questions linked to sustainable development and those linked to collective well-being and quality of life, the hesitations, doubts, uncertainties and questions amount to a paroxysm, as a very broad range of stakeholders and interests to protect must be taken into account.

Based on all of these observations, the problem arises of "de-optimizing" decisions to consider them as a process in which all the hesitations, doubts, uncertainties and questioning that we probe in our daily decisions can find a place. This overhaul of decision-making and the approach to decision support has been systematically developed by the European school of decision-aiding, under the direction of Bernard Roy (who also introduced the verb "de-optimize" in one of his articles in 1968) [ROY 68b]. At the methodological level, the European school for decision-aiding is based on multi-criteria methods that explicitly consider the plurality of points of view considered in a decision problem, technically called criteria. For example, in decisions related to sustainable development, the criteria are the various environmental, social and economic aspects to be taken into account. These underpin any decision we make. The basic idea behind decision-aiding is to provide all the subjects involved in a decision problem with tools that help them to reflect and argue through a model constructed in collaboration with the analyst. In this perspective, to properly understand the message of this book, it is necessary to bear in mind the basic contributions, especially the first ones, of the European school of decision-aiding. Reading through these texts, what leaps out is the effort to give a broad cultural perspective to the different approach proposed. Thus, for example, in Bernard Roy's book from 1985, *Méthodologie multicritère d'aide à la décision* [ROY 85], which has become the reference for decision-aiding, each chapter and each section begins with quotations of some reference researchers such as the chemist Ilya Prigogine, the physicist Bernard D'Espagnat, the philosophers Michèle Serre, Karl Popper and Gaston Bachelard, the sociologists Gregory Bateson, Michel Crozier and Lucien Sfez, each one focusing on the basics of their discipline. In the book *Décider sur plusieurs critères* [SCH 85], from the same year, which thanks to its popular style has certainly provided a fundamental contribution to the spread of decision-aiding, to discuss the cultural bases of the multi-criteria approach, Alain Schärlig refers to, among others, Aristotle, Emmanuel Kant, René Descartes, Thomas Kuhn and Edgar Morin via quantum physics and going as far as Heraclitus. This idea that the European school of decision-aiding bases its scientific approaches to decisions on a broad reflection of its cultural foundations is taken up by Lamia Berrah and Vincent Clivillé. They avoid mere technicality, and rather than listing algorithms, techniques and methods, they retrace, from the earliest eras, the approaches that have been manifested historically to the problem of decisions, with particular attention given to the discussion in the domain of industrial management. Along this journey, Lamia Berrah and Vincent Clivillé rediscover the raison d'être of a decision-aiding approach and provide a perspective that allows us to use with awareness all the algorithms, techniques and methods that the European school of decision-aiding has itself put in place.

Why do we think that the operation proposed by Lamia Berrah and Vincent Clivillé, i.e. a return to the cultural bases of decision-aiding, is important today? To answer this question, it is interesting to refer to an expression that highlights the limits of the optimization approach that the European school of decision-aiding has attempted to surpass. This expression is arithmorphism. It is a concept that was recalled by Alain Schärlig in an article from 1996 [SCH 96] and reprised by Bernard Roy in his article from 2000 [ROY 00]. Alain Schärlig and Bernard Roy present arithmorphism as a tendency to use arithmetic and in general the mathematical approach to express heterogeneous quantities on a single scale, often in monetary terms, as in the case of cost–benefit analysis, in order to aggregate different aspects and to compare each alternative by allocating it a numerical value. This predisposition generates the belief that there is always a decision that is objectively the best, which corresponds to choosing the alternative that has the greatest value. In fact, the concept of arithmorphism was introduced by Nicholas Georgescu-Roegen in his reference work *The Entropy Law and the Economic Process* [GEO 71], with the intention of condemning the construction of mathematical models in economics without the support of a suitable theory, i.e. not supported by a general reflection. We would like to return to the original significance of arithmorphism for Nicholas Georgescu-Roegen as it demonstrates the benefit of cultural reflection in studying questions linked to decisions. Indeed, the problem lies not so much in rejecting the idea that optimization is the sole approach to making decisions, as in rejecting, much more broadly and radically, any non-critical application of algorithms and quantitative methods to decision problems, without considering their articulation with a decision process that aims to construct participants' "preferences" for the decision process, rather than discovering pre-existing preferences in the mind of an abstract decision-maker. In fact, we have to admit that, in general, these discovered "pre-existing preferences" could be very labile and even misleading in the mind of a real decision-maker. Therefore, the problem today, more than 40 years after the birth of the European school of decision-aiding, is that of a "return to arithmorphism", linked to the erroneous belief that the legitimacy of decision-aiding can flow from the mere application of methods that were produced by the European school, as well as by any other technical approach to decision support. To be even clearer, the problem is to reduce the field of decision-aiding to the simple application, for example, of one of the ELECTRE[1] methods proposed by Bernard Roy. It should not be forgotten that these methods, like all other decision-aiding methods, are tools and that to use them well, an awareness is needed that can only be generated by knowing their basic principles and the path that has led to the emergence and acceptance of these fundamental principles.

From this perspective, Lamia Berrah's and Vincent Clivillé's contribution reminds us that reflection on the foundations should never cease, without calling into question the validity of results that can provide us with the tools offered by the school of decision-aiding. We therefore hope that this contribution will meet with the success it deserves and that it will form the starting point for a discussion on the foundations of the decision-aiding approach (and, more generally, of any scientific approach to decision-making), which will be able to give new life to its theory and applications.

January 2023
Salvatore GRECO
University Professor
Department of Economics, University of Catania, Italy

---

1 *ELimination Et Choix Traduisant la REalité* (Elimination and Choice Translating Reality).

Preface

Some Reflections on Decision-making

"Once I've made a decision, I hesitate for a long time". If we choose to introduce this book with the quotation from Renard, it is because it surreptitiously sketches the outlines of the notion of decisions, which is the subject under discussion, while still nevertheless starting to suggest a number of contrasts that could accompany its use: speed and slowness, certainty and doubt, responsibility and misgiving, etc. Thus, this quotation, amusing at first, allows us to touch on an implicit complexity mixed with a sort of apparent simplicity – of this concept of decisions.

Indeed there are "things" that only meet with accurate explanation among the philosophers. Concepts such as beauty, love, freedom and happiness allow each one of us to adopt our own vision, perception or feeling, in the process leaving inaccuracy, ambiguity or indeed error in our speech. On the contrary, the same will not be true of the notion of decisions. The notion of decisions resonates clearly, independently of the level of awareness of those who make them. It is clearly understood. It nevertheless takes on a form of complexity and indeterminism. Also, although the concept of deciding is unanimously understood in the same way and presents a simple and direct definition, the same will not be true of the way in which the decision will be made, a way that is sometimes unpredictable, unsystematic, potentially subjective, unjust or risky, etc. The same will not be true for the form it will take, an answer to a question, a solution to a problem, a reflexive act, one alternative among others, a negation or an affirmation, etc. In more arithmetical terms, deciding could be seen as a function whose domain of

---

For a color version of all the figures in this chapter, see www.iste.co.uk/berrah/decision.zip.

definition would be binary, whereas the resulting decision could take a large number of values.

Clarity has never claimed to be the twin of simplicity. A substantial number of books, treatises and theories have consequently supported the notion of decisions and brought to it clarifications, definitions, mathematizations, methods, tools, etc. The idea of this book is to meet a need of a different nature, that of drawing out the links between the different sub-tonalities and the different parameters of decisions. Just as before for the concepts of objectives and of performance, we seek this time to go beyond the primary simplicity of the concept of decisions and explore its different facets. A systematic look will be adopted with an emphasis on interactions between actors in decisions and the decisions themselves. This will allow us to depart from our usual framework offered by information theory, to visit and survey, with the eyes of a technician on request, the framework offered by the human sciences.

Exploring the outlines of decisions means managing to answer questions such as those listed below.

– What are the parameters involved in decisions?

– In what circumstances do we speak of decisions?

– Do decisions have an opposite, i.e. do non-decisions exist?

– Are there "prerequisites" for decisions?

– Are decisions an isolated and occasional act or an overall and permanent process?

– Can decisions be qualified?

– Can we speak of right decisions and wrong decisions?

– Can decisions be managed?

– What kind of pairing do humans make with decisions?

– How can we explain why two different decision-makers, for a single situation, can make different decisions?

– How do we explain why, depending on the circumstances, a single decision-maker can make different decisions for a single situation? Will the context in which the decision is made impact it?

– Are decisions a thought-through or rather an unconscious act?

– Are decisions accompanied by the tools with which to make them? Do these tools depend on the type of decision to be made?

– What difference is there between a decision, an objective and an action?

– Does the way in which a decision is made impact it?

– Will the object of the decision influence the way in which the decision is made? In particular, do we make the same decisions for ourselves as we do for others?

About 15 questions have been asked, and we expect we will be able to go beyond that. It is these questions that sustain our fascination with the notion and our motivation to study it. Asking such questions and seeing the questions unfold thereby confirms for us the idea that answers given at local and isolated levels will only have meaning in an overall, systematic vision.

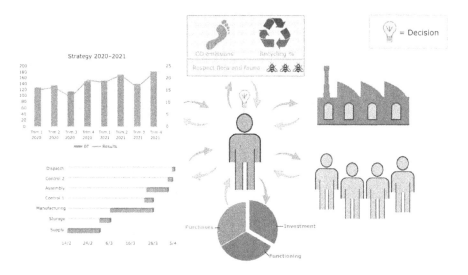

Figure P.1. *Business: a system, a strategy, a plan, a budget, an environmental policy, steering, decisions and people*

The fascination with the outlines of decisions is even more evident when it is applied to a context such as business. Indeed, the business world is renowned for its unfailing use of strategy, precise planning of its needs and a keen steering of its processes, based on the best adapted tools, which are often the most modern (Figure P.1). Nevertheless, humans remain at the helm and remain the main actor in the operation of this system, all the "big" decisions made there belong to humankind, with different ranges as the complexity and diversity of the organization require. Business thus offers the possibility of considering a broad panel of decisions. It will therefore be our preferred foreground for illustrating the process of decision-making.

How do we approach this exploration of decisions? Let us offer the originality of a space for free discourse on the topic. This will at the outset be a search to structure our discussion and foreground the main dimensions of decisions, by classically reviewing the main models for decisions and the main trends of thought.

In addition, let us begin at what seems to be the beginning.

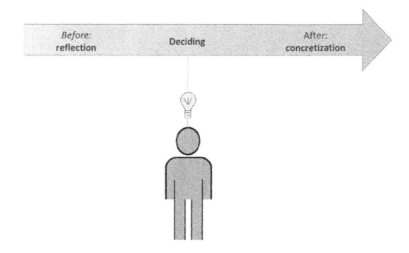

Figure P.2. *The two worlds around the act of deciding*

First, the act of deciding comes to mind. Deciding means "adjudicating", "choosing", "doing what you feel like", "doing what is best", "weighing the for and against", "stopping dithering", "moving on to action", etc. As if, by this act, the being who decides – the decision-maker – moves from one world to another, from the world of reflection to the world of concretization (Figure P.2). As if this act could be therefore, for the decision-maker, the concretization of a feeling, a thought, a reflection, a fear or a desire. This passing from one world to another could be achieved in more than one way.

It could be more or less spontaneous or thought-out, more or less predictable. It could also be more or less bounded by a universe of concretization, which, in turn, could be more constrained. In particular, it could be there to meet an orientation towards a better solution. In which case, this would eventually mean that several scenarios would be available, some more favorable than others. Deciding therefore becomes strongly synonymous with choice. This would then mean that the way in which decisions are made would be identified as a sort of comparison of the solutions offered. We might therefore think that in such cases, deciding becomes more rational and so, perhaps, easier. At least, the problem of deciding becomes the problem of knowing how to choose, and thus of knowing what to renounce.

Thus, we might say that deciding is a boundary between two worlds, a sort of bridge, a sort of passage between them which illustrates, in various forms and under the influence of multiple parameters, some concretization:

– the act of deciding marks the passage from a theoretical/virtual world to a concrete world;

– the passage from a theoretical/virtual world to a concrete world takes various forms.

Let us take care, however, not to identify the first world as that of reflection and thought and the second world as that of matter too quickly. Everything might in fact take place in the world of thought alone.

Two aspects can be identified in addition. The first aspect concerns the interest, motivation, i.e. the underlying intention that would make the decision-maker decide as they do and opt for this or that solution as the case may be. And the second aspect concerns the domain of the possible, in relation to which the decision can be made. Let us pursue both these aspects further.

A person makes a decision with a view to reaching a state, obtaining a situation, whether material or immaterial, hence they decide *for* something. We will call this "something" the objective. Since the objective is the responsibility of its declarant, it will necessarily be impacted by their intention, i.e. their frame of mind in relation to their achievement. This would therefore be the objective, the first element. This may become complicated if the declarant of the objective is not the one who makes the decision. In which case, making a decision would involve a chain – or a loop – in which several actors might be involved. And whoever speaks of actors therefore speaks of potentially different intentions. We might also easily imagine that the intention could then change depending on the "proximity" of the actor to the objective. By proximity, we should understand the link, the expectations and personal projections of the decision-maker on the attainment of the objective. Naturally, the more neutral the link and the less it engages the emotions – the affect – of the decision-maker, the more the act of deciding will be devoid of "passion" and meet "standards". If the decision-maker identifies with their deciding role, it will be different, with the attitude of the decision-maker strongly impacting the result. It is in exactly this context that the famous "charisma" of managers is often mentioned and appreciated. And where this result concerns not only the decision-maker but an organization, the management will take on its meaning and will have the role of weighing and balancing directions. From one thing to another, the "teasing out" around the act of deciding happens very rapidly. For our part, we will often keep the spoken intention of the decision-maker as resulting from their different attributes.

However, could each actor in the chain – or loop – be considered a decision-maker? Or indeed is the only decision-maker the actor involved in the final stage, the one who makes the decision? Here therefore are two dimensions that are beginning to overlap. On the one hand, the intention of the decision-maker conditions how they make decisions. On the other hand, although they are not the only one to contribute to the decision, choices may differ since the stated objective may not be the objective of each or may not mean the same thing for each one. Another dimension could also be imagined, similar to the one we have just described and which would concern the case of several decision-makers. Through their intentions, the decision that emerges might equally represent a compromise situation, just as much as a situation resulting from a more or less disguised "dictatorship" (Figure P.3).

How does the intention impact the decision? What remains of the intention if we are not in a situation to make a choice, or at least, if a person does not feel that they are in a position to make a decision? The answer may be quite clear. In reality, deciding could always be considered as being synonymous with choosing. The variant will appear precisely in a panel of possible choices. There may in fact be offered to the decision-maker a set of well-determined possibilities, even though only a single one may appear to them, in which case the choice will result in a sort of "go" or "no-go", sometimes with one of these two alternatives not even being visible to the decision-maker. "I've no choice! I have to!" or "I've no choice! I can't." Intermediate situations may also happen, in which, for example, the number of possibilities would be unknown, some possibilities would be less clear than others and the only possibility would be imprecise. *Feeling* would replace its characterization. But then, is deciding always relevant in this case?

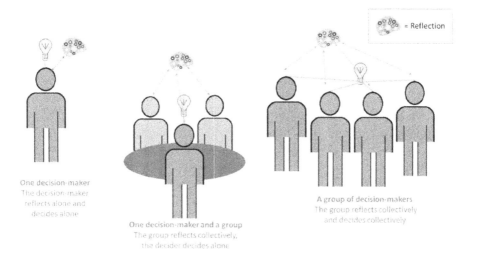

Figure P.3. *A decision-maker, a decision-maker and a group, several decision-makers*

More precisely, must we decide to decide at all costs when we decide to decide or indeed is it necessary to have prerequisites for this? Would these prerequisites involve the decision-maker or indeed the system considered? We could in fact ask about the potential prerequisites for being the decision-maker. Are we decision-makers or do we become decision-makers? How do we declare ourselves to be a decision-maker? Are there conditions necessary to being one? Is deciding innate to humans or is it something that is acquired, a suggestion of humans' environment or humans' contexts? Our reflections may lead us spontaneously to distinguish two situations, one where the decision-maker would be the only actor involved, then impacted by their decision or one where, on the contrary, the decision-maker would not be the only actor involved; where they would decide for a group, an organization or a system. But in the end, does this distinction really have a meaning? In other words, are there situations in which our decision would only impact ourselves, without any effect on the world? No, because we know today that this effect, more or less large, more or less immediate, quite clearly exists. In shorter or longer terms, any act of deciding therefore has its repercussions on a system or organization. However, the outlines of this system or this organization may not be known initially. The nuance will therefore lie rather in the decision-maker having the power to announce themselves as such or in this power being conferred on an entity external to them. The challenge of choosing therefore becomes substantial for the decision-maker. It is without doubt in this sense that recourse to tools known as decision-making tools may be a sort of neutrality and coherence. This may also explain the fact that two decision-makers may not make the same decision. In this sense, the attitude of the decision-maker, mentioned in brief previously, contributes to the ability to be/become a decision-maker, depending on the level of awareness and skills, certainties and fears brought to light in this context of making a decision. There will therefore be a decided and an undecided decision-maker, a confident decision-maker and a decision-maker lacking in confidence, an optimistic decision-maker and a pessimistic decision-maker, a sincere decision-maker and a manipulative decision-maker, etc.

At the same time, is it possible, in life, to abstain from deciding? In other words, can we declare ourselves a non-decision-maker (in absolute terms)? Envisaging an eventuality and its counterpart allows us to consider the existence of contexts (systems, situations, moments) for which we can declare ourselves a decision-maker and situations for which, on the contrary, it would be better to abstain from deciding. The extreme case is one in which a person makes the decision not to decide. Non-decision is consequently a decision. Naturally, non-decision should not be confused with indecision, even though the latter may lead to it. This capacity for discernment will enable "bad" decisions to be avoided. Moreover, it will allow everyone to be in their right place, and will allow everyone the potential to "become" a decision-maker. Figure P.4 illustrates these respective situations of decision-making, indecision and non-decision for the particular case for which, faced with the multiple possibilities offered, deciding amounts to choosing one of these possibilities.

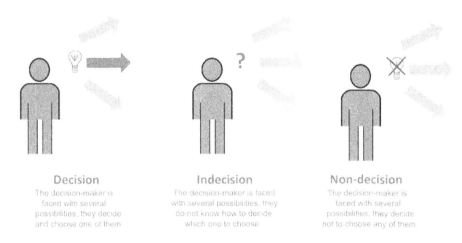

Figure P.4. *Decision-maker, decision, indecision and non-decision*

At this level, for which we like the mathematical formalization, fuzzy sets theory might be of help. This theory, which has already been tested for the declaration of objectives and expression of performance, can in fact be presented to capture subjectivity, uncertainty, imprecision, doubts, indecision, etc. As many parameters as there are known attributes of the act of deciding.

Let us now consider some preliminaries to the act of being a decision-maker (Figure P.5):

– behind every decision is a decision-maker;

– the act of being a decision-maker is not something decreed but something lived;

– the quality of the decision can be impacted by the decision-maker's intention;

– deciding involves the attitude of the decision-maker;

– a decision-maker decides in an adequate context.

Naturally, when reading these properties, we can imagine AI devices (artificial intelligence) and the potentialities it could offer on the subject of relatively automated, robotized decisions. This situation will escape our discussion, highlighting, from our point of view, the limits of the decision made in the classical sense, so by a decision-maker, and covering a fairly restricted space for decisions.

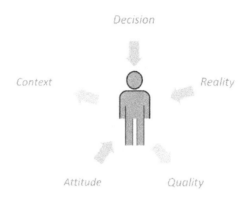

Figure P.5. *The decision-maker and decision*

We finish this aside on the revelation of the decision-maker and resume our reasoning. When we decide, we decide to cross a boundary with a view to an objective to be reached. This objective will certainly be that of the decision-maker, but it will have a range, it will involve a system, a variable or a quantity in relation to which the objective will be declared, the decision made and the result evaluated. "Something" will therefore be decided. We would like to think, as engineers and as information and modeling scientists, that the more that "something" is linked to an objective, material, easily quantifiable quantity, the more that objective will be absolute and shared. But this would not take account of the affect of the decision-maker, as mentioned above, which broadly conditions their intention and attitude. It is in this register that we might dissociate situations from desire, from a pure vision by the decision-maker, from those of insufficiencies or gaps to fill, an objective and rational observation. Deciding will not happen in the same state of mind; just as decision will not have the same qualifications (importance, risk, etc.). We might thus imagine the following properties:

– the decision involves an objective to be attained;

– the nature of the objective conditions the decision.

Let us go over the act of deciding once more. Certainly, we decide "for something", but also "from something" and "in relation to something". The "from something" is attached to the act of deciding, and the "in relation to something" is linked to the system. The system will offer the variables in relation to which the objectives will be stated. We can in fact imagine the singularity as well as the plurality of these variables (and objectives). Moreover, it is according to these variable(s) that the decisions will be characterized as big, important, minimal, without risk, etc. There too, decisions will not be made in the same way. And this time, these qualifiers will link the act of deciding to the system. At this level, we note that, naturally, the more the decision-maker "knows" the system, the clearer the act of deciding can be for them. But as any system exists in an environment, this knowledge and clarity are often joined by uncertainty.

The following properties also sum up these remarks:

– the decision involves a system;

– the nature of the system variables which are involved in the decision conditions the decision;

– knowledge of the system and its environment impacts the decision;

– decision and uncertainty may be mutual.

Let us return a while to this notion of the system's environment. This environment will identify part of the context in which the decision – in relation to the system considered – will be made. Stating the hypothesis that the decision-maker will necessarily be an observer of this system, we will link the decision-making context and the system's environment. Part of this environment will be known, clear and observable. Another part will be implicit and unaware. We might therefore speak of culture, conditioning, rules of life, of habits, customs and traditions. These two parts would explain how the decisions are made. The company, which is a conventional and highly formalized system, will not escape the rule of these two parts. Additionally, part of the impact of decisions on such a system could be quantifiable, whereas others could be felt, and in some cases even unexpected.

Moreover, let us establish at this level of our discussion a working framework of a system external to the decision-maker (Figure P.6). We will leave aside decision situations for which the decision-maker is led to decide for themselves. We will focus instead on those situations where someone makes a decision for a system that has material existence.

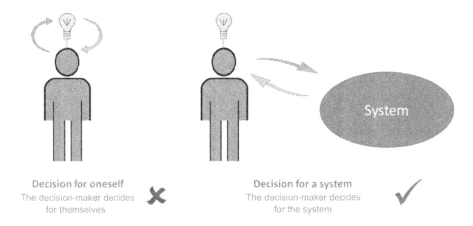

Figure P.6. *A decision by the decision-maker for a system*

So more generally, we can consider that four parameters constitute the act of deciding (Figure P.7):

– the decision-maker, they who decide, their skills and knowledge, their affect, their intention and their attitude;

– the object of the decision, the "something" in relation to which a decision is made;

– the objective of the decision, the "for something" that motivates the decision;

– the decision, as such, the "from something" that results from the act of deciding.

From this first structure we can, as we began to do earlier in this discussion, imagine all the variants of the act of deciding, depending on the attitude of the decision-maker, for example the importance and nature of the object of the decision or the objective, etc. These are therefore variants that mean that deciding fits more or less into the chain of "reflecting, discussing, organizing and, finally, choosing", whether it is an immediate act or rather an act that is prepared, an act of certainty or one marked with uncertainty, fear or risk, an objective act or, on the contrary, an emotional reaction. The decision-maker, the object of the decision, the objective of the decision and deciding become consubstantial facets of the act of deciding.

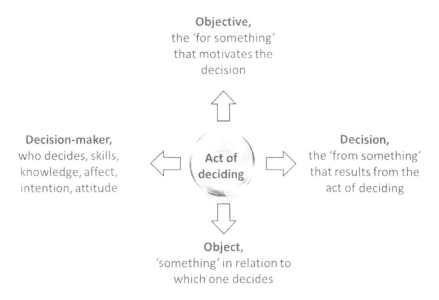

Figure P.7. *The four facets of the act of deciding*

This thus leads us to seek to position the act of deciding, on the one hand, and the decision itself, on the other hand. Intuitively and this is what we supposed, we would like to say that the decision is the result of the act of deciding. To go in this direction, we note that the decision is "made". It cannot be owned or obtained. This indeed reinforces the idea that it is the result of a form both of projection or reflection, and of action, a concretization of the projection or reflection. Deciding is the action, whereas the decision is its result. The decision thus becomes the concretization of a process, of this passage between two worlds – the theoretical and the concrete. This process – of making a decision strictly speaking – may be more or less complex and involve one or more variables. By processes we understand, usually, the succession of a number of stages, which will allow the transformation of an idea, a feeling, an objective or an observation into a decision. The process will be the structuring model for the chain mentioned previously. The large bricks of such a process can be distinguished intuitively according to three stages, which would correspond respectively (see also Figure P.8) to:

– everything that is prerequisite to the decision;

– the decision;

– what will be a corollary to the decision.

The "before" of the decision will concern all the prerequisites needed to make the decision, i.e. the identification (perhaps both) of the decision-maker, of the object of the decision as well as its objective. The decision will be the orientation given. It remains a mystery to us for the time being. The "after" of the decision will influence the implementation of the decision made and potentially the expression of an associated performance in view of the results obtained. Naturally, this naive characterization will have a strong need to rely on those offered by the literature that we will encounter in the following chapters.

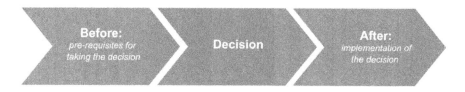

Figure P.8. *A naive vision of the process linked to the act of deciding*

It will be the steering of these processes that will almost immediately lead to the right decision being made; the thinking will move, for industrial engineers, towards the precepts of the Deming wheel (Plan, Do, Check, Act) and the positioning of the decision in its first and fourth respective stages (Figure P.9). The decision will have its place in the planning stage (*Plan*), the one that concerns the "what to do" and the "how to do it"; of which a very specific case would be the binary situation mentioned above ("go" or "no-go" amounts in fact to "do that" or "do not do that"). In the case of a negative vision, the decision is not to carry out the envisaged objective; no planning will therefore be forecast. Although the *Do* and *Check* stages concern respectively the execution of the planned actions and the expression of the associated performance, the *Act* stage will naturally concern the decision too, since the Deming wheel is essentially cyclical. It is at this level that discussion around the notion of reaction can be envisaged. A reactive decision can mean a form of haste that would lead the decision-maker to decide poorly. A reactive decision can equally mean a decision that follows an analysis and an observation. We will repeat ourselves once again: the consequences will differ depending on whether the reaction concerns the system or the decision-maker.

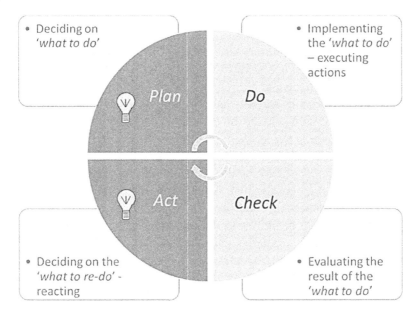

Figure P.9. *The decision in the Deming wheel*

We could, however, enquire about the input data for the decision process. Although we are still inspired by the philosophies of continuous improvement, these data can identify an existent (*as is*) that allows an objective (*to be*) to be declared. A diagnostic stage, involving expression of needs, will therefore be required for the decision. In cases where desire and intuition are involved, the diagnostic stage will not be needed. Outside this particular case, a minimum of explicit information is therefore opportune for making the decision.

But finally, how does the decision concretize? Spontaneously and in view of our analysis, we would think of a transition to the act, to an action. The decision becomes the artifact that very closely precedes the action. The action materializes the decision, and allows an objective to be reached, linked to the act of deciding. Deciding thus becomes a process of which the result is a decision, or an action to be carried out. By action, we naturally mean the concept, which could take the form of a set of actions. It is therefore the action that materializes and gives an existence to the decision. In other words, we find again the notions of intention, or indeed a wish or a hope. This action could just as well have a reality in the physical world as not, at least in the immediate sense.

Moreover, if the decision is clear, do we still talk about a decision? We would like to think that we seek to decide when we do not know how to decide, so that the decision, the choice or the action is not clear to the decision-maker, who consequently has a need to define a path for themselves to lead to this decision, choice or action. At this level, the challenge of constructing this path is posed. Will it be the product of the decision-maker's cogitations, of a flash of intuition or reasoning based on appropriate methods and tools? The process of making the decision therefore becomes an assisted one. There is thus the decision that the decision-maker makes without being aware of it. The "something" and the "for something" would in this case be linked to physiological quantities leading to automatic acts. They could also be attached to simple, mastered and routine acts without any equivocal meaning. On the contrary, there is the decision that the decision-maker makes in full awareness, and potentially takes time in making. Such decisions would perhaps be unique, unusual, unpredictable and complex. The process of making a decision would then be seen sometimes as almost automatic and sometimes subject to reflections and hesitations. Not knowing how to decide amounts to meaning that the decision is made not reflexively or without awareness, but that it requires reflections and the crossing of a bridge.

We should nevertheless be careful about this notion of time. We are afraid of the deduction: "I'll decide right away whether it is simple to decide". Certainly, the time factor could be involved in one situation or the other. But we can equally well imagine being in a context where the decision may be objectively simple but the character of the decision-maker – hesitation, caution, doubt – causes them to take time before deciding on a "something" that is clear. We spoke of indecision. The obvious character of the "something" therefore becomes not absolute but relative to each decision-maker, who will not only decide depending on who they are but also depending on their intention in this act of deciding. Just as there may be complex situations for which the decision processes can move very quickly, so long as the elements to consider are clear to the decision-maker. Thus, we would like to check whether the object and the intention attached to the decision are clear to the decision-maker and whether the latter presents an "objective" attitude in this context; the duration of the pathway to the decision will be the one necessary for the decision to be made. This duration will be that of the cogitation and analysis, whether assisted or not. Now, there is the intuitive decision, certainly the most well-founded from our point of view, but which requires the decision-maker's capacity to access their intuition. This is another debate.

In the matter of decisions, there would therefore be the simple and immediate situation and the complex and delayed situation. Would the typology of decisions be binary like this? Indeed, since a decision belongs to the decision-maker, it will depend on their feelings in relation to their ability to make it. In the case where the decision-maker is not able to decide, this might equally be due to their attitude or more objectively to a lack of elements or information necessary to make this decision. The attitude of the decision-maker is certainly involved in making the decision (and we have discussed this broadly until now), but this attitude is supplied by information on the object of the decision, indeed of their objective. In relation to this aspect, the way of presenting this information can more or less sway the decision one way or the other. If the table presented is attractive for a "go-getter" decision-maker, the decision will be predictable. Similarly, if the scorecard presented is worrying, the decision of a careful decision-maker will be equally careful. In addition, if it can be subject to interpretation, the information will not have the same effects on the decision-makers. The context and environment of decision-making could be illustrated via these situations. Two different decision-makers would draw potentially different conclusions from identical scorecards. Figures, emoticons and other precepts of visual management could aid the neutrality and commensurability of the information involved. Tools for aiding and analyzing the decision will be, too.

Information and scorecard in a decision context: this is what makes us turn systematically on the notion of the expression of the performance. Can a decision be evaluated? Let us return to our discussion of "right" and "wrong" decisions. Traditionally, the performance indicator methodologically links a variable of the system, an objective and a measurement linked to reaching this objective. Formally, the performance indicator will "calculate" a piece of information (the expression of the performance) on the basis of a comparison between the objective (the expected state) and the measurement (the achieved state). The "better" the expression performance, the better the objective attained. Does this still mean that the decision made in this sense is the right one? A right decision would thus be a decision that would allow the objective to be reached. And who decides this? Figure P.10 shows an example of a "right" decision, attested by the increase in value of the OEE (Overall Equipment Effectiveness), the well-known indicator for the yield of equipment in workshops.

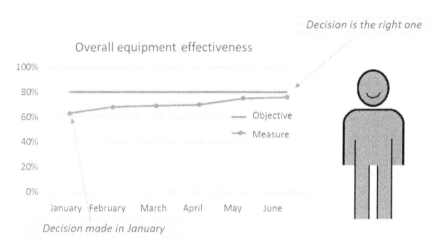

Figure P.10. *The decision, the performance and the objective*

We have now arrived at the end of our reflections on the notion of decisions. We are convinced without doubt of certain clear facts about decision: its systemic character, its environment, its process, the role of the decision-maker, the parameters it involves as well as the diversity of forms it can take. Strengthened by this impregnation and the main keywords inherent to the decision (Figure P.11), we can now better understand the main currents of thought in decision theory, the subject of the following chapters of this book.

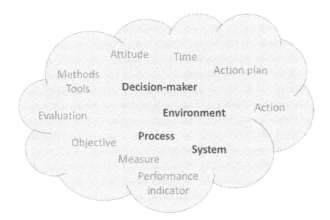

Figure P.11. *Keywords on decisions*

More precisely, we begin (Chapter 1) with a broad exploration of the notion of the decision environment. We then address (Chapter 2) the notion of the decision process that we gladly confuse with the notion of the decision and the act of deciding. Schools of thought dedicated to the making of decisions are then considered (Chapter 3) before finally adopting a data processing point of view and seeking to position (Chapter 4) some "big" methods to aid this decision in the landscape. Throughout this book, we will not stop analyzing the impact of our remarks on the decision-maker/decision pairing, where our terrain of action is the industrial environment.

January 2023

1

Decision and Decision Context

Ultimately, when a decision-maker makes a decision, it is with the intention that this decision, whatever it is, brings something better to the situation considered, since, in any case: "The essential thing is not to live, but to live well."[1]

1.1. Introduction

The reflections made previously allowed us to really sense the major elements linked to the decision. The concepts underpinned by the decision are in fact numerous, and their interactions make it into a consubstantial whole.

At the very least, the decision will be tied to its proximity to the decision-maker, on the one hand – the main focus of our study – and its systemic dimension, on the other hand.

Indeed, some keywords emerged from this quasi-intuitive analysis, concerning in this case: the intention and objective of the decision-maker, the system and the variables to which the decision is attached, the prerequisites for the decision as well as its process and its potential evaluation. There are as many keywords as questions.

The essential pairing is the decision-maker/decision pairing. Beyond this, the object and objective of the decision are the two other structuring

---

[1] Translation from a quotation from Plato on the thoughts of Socrates [BRI 05].

parameters of this history (Figure P.7). And, indirectly, the decision-making context comes into play; this "set of circumstances in which an event occurs, in which an action is situated" [LAR 21]. The context is this parameter, so broad that it covers all aspects, which means that the decision, for a decision-maker and a system, is not made in the same way. The principles of system modeling allow us to hide behind the notion of the system environment to approach this notion of context. Thus, as mentioned before, we will assimilate the decision-making context and the system's environment, in relation to which the decision is made. Indeed, the system's environment, including its observer and our decision-maker, could be confused with the latter. We can agree that the context in which the decision is made is the environment to which the decision-maker and the system are respectively attached. The subtle question of the boundary between a system and its environment will nevertheless escape our discussion.

The context of decision-making, or of the system's environment, thus identifies all the aspects – material or immaterial – that, through interactions with the system or the decision-maker, have an impact on the decision.

To discern this impact, we propose to identify this environment with the different living environments in which humans have become accustomed or been conditioned to make their decisions. We choose to "think big" to do this and to grasp, through the "living environment" the various societies, cities and civilizations built and lived in by humans. We are all convinced of the fact that some rules, beliefs and visions of life are greatly dictated by the cultural and religious aspects of societies, for the whole domain. In reality, we are aware, even beyond this, that our observations will not so much concern the system as the decision-maker themselves. In the decision, it is the decision-maker who makes the decision for the system and not the system that decides for itself. This tautology therefore allows us to distance changes – the events – produced independently of the objectives of the decision-makers.

More precisely, the "quest" of this chapter will be to answer the following questions:

– What elements may have influenced humans in the decision-making mechanism?

– How is this influence manifested in this mechanism?

To support our discussion, we recall the etymology of the word "decision". This word comes from the Latin *decisio*, and means the "result of reflection making it possible to solve a problem". From this definition alone, we see a vision centered on the need to find a solution to a problem through reflection. Here is an anchoring of the decision centered on the problem/solution pairing. Is this anchoring universal, with an absolute perception? Or does it belong to Western society and culture?[2] Is it necessary to envisage a problem before deciding? Is reflection the only means of deciding? In this chapter, we leave aside ontological considerations of the concept of decision. Instead of the questions/answers practiced above, we favor an observation/analysis of the societies and cultures in which decisions were/are made.

Given the possible scope of this perspective, we propose to put forward some assumptions of analysis as well as some restrictions for the perimeter of study. The simplest angle for doing this seems to us to be a chronological one. This angle is, moreover, one that fits the vocation of systems of being in a constant state of transformation over time. Our approach will remain global, targeted and centered on the founding elements of civilizations, on the one hand, and their defining elements, on the other hand. It is naturally within this vision, in which a decision-maker decides for a system other than their own (Figure P.6), that we place this analysis. Moreover, to avoid losing ourselves in the different possible points of view (culture, society, city, etc.), we choose to rely on both – complementary – definitions of the term "civilization" given by the CNRTL (*Centre national de ressources textuelles et lexicales* – National Center of Textual and Lexical Resources):

– "a more or less stable (durable) state of a society that, having left the state of nature, has acquired a high degree of development";

---

2 A Chinese master explains: "Certainly, these are difficult notions for you to grasp. What do our texts translated into western languages give us? Just as we translated the Buddhist sutras, you must have found western terms that surely don't fit the text perfectly and lead you into error. Your concepts result from Greek philosophy and Christianity. I imagine that borrowing from them, for the translation of our texts, must dress them up in strange costumes! Would it not be preferable to leave them in phonetic transcription to retain their autonomy; to stop explaining them, with the help of different passages where they are cited, to provide them with at least one approach? You would enrich your vocabulary with words such as Tao, Li, etc. Tao is neither your God, nor Being, nor a principle that governs the universe, but may be a little of all these things. Li is not what you call reason, or logic, but is not completely unfamiliar" [VER 03].

– "a transmissible set of values (intellectual, spiritual, artistic) and scientific knowledge or technical realizations that characterize one stage of progress of a society in evolution".

The idea behind this chapter is therefore to describe these durable states of societies. This description will make it possible to go on to meet different contexts, to observe, for each, information, values and knowledge, etc., as many prerequisites as necessary for the decision (Figure 1.1).

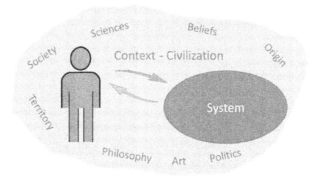

Figure 1.1. *Analytical framework. For a color version of this figure, see www.iste.co.uk/berrah/decision.zip*

During this review, we will sometimes depart from this approach. We will focus at times on quotations from some philosophers, scientists and other thinkers, given that these quotes can help illuminate the notion of decisions. These individuals will be from Greek Antiquity, will have lived during the era of the Arab–Muslim miracle, lived during the Renaissance era, or that of The Lumières... They will also be able to convey the wisdom of Asiatic thought.

We will therefore traverse centuries and millennia... without doubt clumsily, perhaps stealthily or surreptitiously, taking care not to get lost and seeking to seize, deduce and understand the impact of context on the making of decisions. Naturally, the avid reader of philosophical, mathematical, scientific, etc., knowledge about the notion of decisions is invited to consult references such as [ISE 93; BER 13; BOU 06a; BOU 06b; BOU 06c; MER 16; TEI 19]. For our part, we will concentrate on a historically explained perception and then on a systematic consideration of the latter... Let us set out on a whirlwind trip through time!

1.2. A fleeting look at some of the great civilizations

1.2.1. *Method of investigation*

We have chosen to "go through" history. This journey will be led by some references on the subject, in this case [DUR 35; CRO 63; BRA 13; DEM 18; LOP 12], but it will naturally be imbued with some subjectivity. Nevertheless, our approach can be summarized in the following lines. We have begun by choosing the most ancient civilizations in history. We have therefore allowed ourselves to wander through time – at the mercy of our knowledge and some proximity to our own context – to arrive at today's context. Because of this, we are aware that many civilizations, such as pre-Columbian, Oceanian or African ones, have equally reached high levels of development and transmitted values and knowledge to all humanity. Leaving their study to future investigations, we still make the hypothesis of a form of similarity between societies, and of the functioning of humankind, across time and space, through universality and atemporality[3].

The timeline presented in Figure 1.2 positions the considered civilizations chronologically. Given that there is generally little consensus on a civilization's start and end dates, the dates proposed will fit simply with its "high point" or "golden age", i.e. to take up one of the two previous definitions, its stable state around a high level of development. So, for the Western world, for example, the start will be situated at the Renaissance, the period that marked a strong break from the medieval world. For some civilizations such as that of Ancient Egypt, the end date corresponds to the first signs that the functioning of the Pharaohs was declining. In line with general understanding, we have chosen to keep a single definition for some civilizations, even though their golden age may have been several centuries ago. India still remains India.

A description – summary – of each civilization chosen is given in conformity with what we declared previously, i.e. for a considered civilization:

– its origin and geographical location;

– its beliefs, religion and philosophy;

---

3 "Humans are humans everywhere. Otherwise, how could we understand each other? Only a unique history separates us" [VER 03].

– its social and political organization;

– its various artistic and scientific developments;

– some thoughts that reveal the decision-making context.

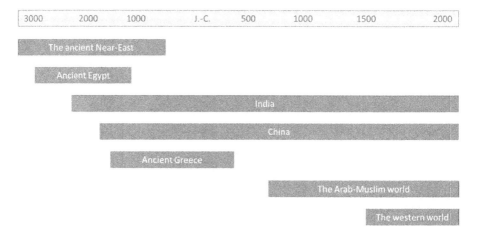

Figure 1.2. *Timeline of the civilizations chosen. For a color version of this figure, see www.iste.co.uk/berrah/decision.zip*

We will not fail, throughout this description, to take time out to interpret and to propose anything that we consider essential for our analysis.

We will also point out that, depending on the information we have been able to equip ourselves with, the dates of events belonging to the civilizations considered may be more or less exact, ranging from millennia to years, or centuries. Since the precision of the dates is relative in our study, we assume this form of heterogeneity in the description.

It is said that "history[4] begins at Sumer". We therefore start with this first civilization born in the mysterious Near-East.

1.2.2. *The ancient Near-East*

The ancient Near-East, the writing, the legends, the thought... Around 3,300 BCE, a people coming from the West settled in a territory then called

---

4 History is in fact defined as the period of humanity accessible through writing [MAR 96].

Sumer and inherited its name. Situated between the Tigris and the Euphrates, "the land between the rivers", the Greek translation of which was later "Mesopotamia" [BOR 14], saw the Sumerians prosper, until take-over by their Akkadian neighbors around 2,200 BCE From 1,800 BCE, the ascension of the Assyrian people, coming from the north, began. This people progressively extended its influence beyond Sumer, an influence that ended in 612 BCE with take-over by the Chaldeans. The latter were rapidly replaced in their turn, around 539 BCE, by the Persians, whose empire to the east was greatly expanding. On the periphery of this empire, Mesopotamia then became one of its provinces and saw its level of development decline. Many people thus lived in the region, people who dominated it like those mentioned previously, as well as peoples who lived there – for shorter or longer periods – such as the Arameans, the Hebrews, the Kassites, the Hittites, the Arabs, etc. This doubtlessly explains the feeling of greater or lesser belonging that each had to this land. The ancient Near-East was hence born in Sumer and declined 3,000 years later, to be subsumed into the Persian empire.

TO REMEMBER.– Humans have been continually conditioned to *fight to obtain*, possess and conquer the object, *then protect* what they have conquered. A *repetitive* decision model based on an *objective* of *expanding* and then in turn *perpetuating* the system. An environment of *competition* and risk, an action plan in turn offensive and then defensive, have been resonant since that time.

The Sumerians had an uneasy relationship with nature, fearing phenomena they could not control. To help themselves understand nature better and to allay their fears, they adopted a set of beliefs linked to the existence of multiple gods, steeped in the belief that these were the gods who made the world function, and who created both the good fortune and the misfortune of humans, depending on the level of tribute paid to them.

TO REMEMBER.– An *explanation of the decision model*: as soon as the system is under control, a decision is made to expand it; where it is not under control, beliefs were used as a form of reassurance and as a guide in the matter of protective actions to be taken. Because of this, a *decision* was only *made* within the *framework of these beliefs*. Thus, humans were convinced of their own powerlessness (it was the gods who made decisions) and of their culpability in relation to what could have been done (homage to the gods) to avoid what happened (natural disasters). The creation of guilt and powerlessness – rather than the creation of responsibility – with fear as a foundation. *Limited decision-making in an uncertain environment* was therefore contemporary with Sumer.

However, from 1,200 BCE during the reign of the Assyrians, a chief god emerged, Marduk. The practice of questioning then began to develop, along with views of life events that had a more spiritual meaning[5].

At the beginning, society was organized around cities that were independent and expanding, despite economic and military competition, bringing together several thousand people. This intra-territory competition ended with unification by the Akkadians and the appearance of a centralized state with a king at its head acting in the name of the gods and a capital. The state acquired a hierarchy with, at the highest level, the nobles who included ministers, governors, stewards and relatives of the king, priests authorized to worship the gods in the temples and possessors of large estates. At the middle level were freemen, peasants, fishermen, artisans or scribes. At the lowest level, slaves were considered chattels that could be bought and sold at will. The state was endowed with a stable political and social organization that relied on texts and laws, including the famous Hammurabi code and its decisions on justice, written in Babylon around 1,750 BCE.

TO REMEMBER.– Here is a *framework*, built on *several levels*, that made it possible to make decisions and to identify the perimeters of each decision-maker in the system. *Centralization* and the notion of a *head*, *of a plural decision* and *formalization* were already on the menu.

Moreover, this civilization saw humans make inventions in numerous domains, primarily writing, or using reeds to draw pictograms on clay tablets. Writing made it possible to record private actions such as inventories of goods, business transactions, law codes and also the epics of historical or legendary individuals or even religious rituals. In parallel, the creation of irrigation channels and the invention of the metal plowshare facilitated agriculture and transformed the territory into a "fertile crescent". Other inventions involved the development of craftsmanship using the wheel,

---

5 Recall that the Ancient Near-East was also the cradle of the monotheistic religions – Judaism, Christianity and Islam – of their prophets. Since this issue is not only a historical one, we leave to the experts its analysis.

weaving and metal working. The ancient Near-East established commerce within its territory based on bartering. By means of caravans, this commerce extended gradually to different peoples, either neighboring or more distant ones, in Egypt and in Asia. Moreover, new architectural techniques made it possible to found Babylon and its hanging gardens or even to build temples, including the Tower of Babel, an attempt by humans to reach the gods. The scientific dimension was also present, in the practice of astronomic observation, the definition of a lunar calendar and the establishment of the sexagesimal counting system. In this fashion, "ancient scholars, from the first half of the second millennium at the latest, had 'in their way' and according to their reasoning, discovered abstract thought, analysis, deduction, research and the establishment of principles and laws: in a word the essence of the method and the mindset of science" [BOT 97].

TO REMEMBER.– Humans of the ancient Near-East had not ceased to grow and perhaps to seek to improve their quality of life, and did so by *growing their system*. A line of action which was dedicated to the purpose of making humankind's system durable.

And to finish, let us cite two quotations contemporary with this era:

> When it is about some-one else's bread, it is easy to say: "I will give it to you", but the time of actual giving can be as far away as the sky[6].

> Be not too sweet, lest they swallow thee, be not too bitter, lest they spit thee out[7].

TO REMEMBER.– Here in the first quote is evidence of a possible *shift* between a decision-maker's *intention* and their *actions*. The decision may not reflect the intention because the intention is far removed from the decision, this famous *process* that we will visit in the next chapter. As for the

---

[6] Quotation taken from the French translation of the work *The Instructions of Shuruppak* [ALS 74]. Shuruppak was a Sumerian king reigning approximately 2,500 BCE; https://etcsl.orinst.ox.ac.uk/section5/tr561.htm.

[7] Quotation taken from the French translation of the work *The Story of Ahiqar* [NAU 09]. Ahiqar was a sage living at the court of an Assyrian king, approximately 680–660 BCE; http://www.syriacstudies.com/AFSS/Syriac_Articles_in_English/Entries/2007/12/12_THE_WOR DS_OF_AHIQAR_Aramaic_Proverbs_and_Precepts_-_Translator__H._L._Ginsberg.html.

second quote, it reminds any decision-maker to guard against extreme behavior and its consequences.

1.2.3. *Ancient Egypt*

Ancient Egypt, the gift of the Nile, the magnificence, the adoration... A little before 3,000 BCE, powerful Upper Egypt imposed unification on Lower Egypt, close to the fertile Nile delta. From this unification (ancient), Egypt was born; a civilization that lasted 2,000 years in near-total stability, as much on a territorial as on a populational level. At its height, between the 15th and 10th centuries BCE, Egypt extended over a territory ranging from Nubia (modern-day Sudan) to Syria. Egypt was seen by its neighbors as the model to imitate because of its economic, political, religious and artistic perfection. This interest, at first contained militarily, led Ancient Egypt to lose its grandeur little by little, as it experienced mounting defeats by Libyan, Nubian, Assyrian, Persian and Greek neighbors, until it became a Roman province in 30 BCE.

TO REMEMBER.– A system that had known 2,000 years of *stability*, expansion and resistance to environmental threat, a form of durability. Humans knew how to make its system last, to make it function and to transform it in this sense. How? Rather than by power and force, this happened through a coherent and reinforced identity and alignment. *Constancy in its intentions and alignment between objectives, decisions and actions* had been the governing strategy of the Egyptian system.

The Ancient Egyptians believed very strongly in a possible extension of life beyond death and in an environment beyond the perceptible: the universe, of which they were part. They had a complex vision of the constitution of humans and believed in a force, a vital energy and spiritual double – the *Ka* – that linked them to the universe and on the harmony of which life and immortality depended. These beliefs, marked by the fear of disappearing and the quest for immortality and the magnificence supposedly associated with the latter, had led Ancient Egypt – too – to seek continually to understand the universe. This suggested a dual vision to them. Good and evil, light and darkness, day and night, order and chaos, perfection and completeness, immortality and the shadows, etc., thus structured thought, myths, the gods and actions. As a case in point, we find Ma'at, a goddess who represented good, harmony and justice, and her opposite Ifset who represented demons, chaos and injustice. In the same way, Osiris was the god of world order, while Seth was the god of disorder and disruption. Central, the death myth conveyed the idea of the conditional accessibility – depending on the dignity of the soul – of a life after death. This life signified eternal rest, with the gods. Fear thus continued to influence humans, alongside a search for growth, indeed perfection and eternity.

TO REMEMBER.– *Understand to master an open system with undefined outlines.* Adopt a bipolar *vision of any element*; through what can be *beneficial/good* and what is *not beneficial/bad* and in a corollary decision. Implicitly, judgment and e*valuation* will lie upstream of any reflection. Is this not our vision in the stages of our analysis, those that lead us to weigh the "for" and "against" before deciding? Performance is synonymous with perfection. A *simple*, *direct* approach aiming for the o*ptimal*, conditioned nonetheless by a bipolar modeling of the system.

Outside the role they hold in the founding myths of Egyptian religion (relating respectively to the creation of the world, to life after death and to the cycle of day and night [SCH 03]), the gods formed a system. Existing in a hierarchy, distinctions were made between primary and secondary gods, the latter subordinate to the former. The main gods personified natural phenomena or related to human feelings and activities.

TO REMEMBER.– *Everything has its place and everything has its role in a whole formed of things and their opposites.*

The valley of the Nile was the heart of ancient Egypt, then structured into several provinces. Power was centralized and exercised by the Pharaoh and god together. Society was hierarchical with, at its top, the Pharaoh, assisted by the grand vizier and the governors of the provinces, for all day-to-day business. Then came the ruling class, which included the nobles, priests and scribes and finally the "people", which included the peasants, artisans, soldiers and servants. The nobles possessed large agricultural domains on which they put their peasants to work. For their part, the scribes were in charge of administrative formalities, systematizing information processing, which gave rise to the creation of the ancestor of the information system. Moreover, exchanges were developed: internal, via the riverways, and external, via sea routes, this time via the Mediterranean, the Red Sea or the

Indian Ocean, in the direction of Sumer, the east coast of Africa or even India.

TO REMEMBER.– *Information formalization* and *processing* are two key terms that we can retain from this civilization.

Ancient Egypt lay at the origin of many inventions. Thus, a little time after Sumer, it invented figurative writing formed of hieroglyphs, which was used for economic, administrative and religious activities. The yield from agricultural lands increased thanks to an irrigation system being put in place based on the "shadoof", a device that made it possible to raise water drawn from along the Nile by means of a see-saw mechanism. Artisanship also developed with the caulking of ships, the discovery of glass and Egyptian faience using the firing of silica. Moreover, Ancient Egypt built religious temples and pyramids and the tombs of the Pharaohs, including sculptures, decorative frescoes and furnishings following a precise code. The Ancient Egyptians were also great scientists. "Don't be proud of your knowledge. Consult the ignorant and the wise; The limits of art are not reached, No artist's skills are perfect; Good speech is more hidden than greenstone, Yet may be found among maids at the grindstones"[8]. Egyptian medicine opened the way for "specialization" and developed surgery. In mathematics, Ancient Egypt used the decimal system (since zero did not exist, the numbers were formed by juxtaposing hieroglyphs for units, tens and hundreds, etc.), mastering arithmetic and developing algebra and geometry.

1.2.4. *India*

India, Buddha, interiority, Vedism, the caste system… Around 2,500 BCE, nomadic peoples descending from the north became established in the Indus Valley to the north west of modern day India. Later, around 1,800 BCE, the Aryans (coming from Iran) mingled with indigenous populations and settled on the plain of the Ganges, to the north east of modern day India. In the 4th century BCE, an indigenous people, the Maurya, unified the Indus valley and the plain of the Ganges to form a territory extending over the Indian sub-continent (India). From the 4th century until the middle of the 20th century, through the successive arrival of Huns, Arabs, Afghans, Mongols and Persians and finally Western colonization, India was divided and subject

---

8 Quotation from Ptahotep, vizier c. 2,400 BCE; https://www.goodreads.com/quotes/9701977-the-instruction-of-ptahhotep-instruction-of-the-mayor-of-the.

to the influence of these respective invasions. It achieved unity and independence in the middle of the 20th century. Indian civilization was thus born in the Indus valley and maintained for more than 4,000 years over a vast territory, divided many times by invasions that ultimately were always pushed back.

TO REMEMBER.– *Two* profiles of decision-makers: those motivated by a conquest *external* to their territory and those motivated by conquest *internal* to their territory alone.

In fact, the spiritual dimension is strongly present in India[9]. The idea that life is only a stage or a means for "something else" is present there, as in Ancient Egypt. But rather than eternal rest for the soul, it is this time liberation, a liberation of each from on high, of the burden of their thoughts and actions. Liberation is synonymous with death, definitive or temporary death depending on our understanding. Ultimate death is synonymous with fusion with an absolute, ideal whole. And it is by means of a quest, which may be more or less progressive and quick, for truth and knowledge that this liberation will happen. What happens outside (of the self) is a reflection of what happens inside (of the self). We will leave aside evaluations of what might be "good" and what might be "bad" to seek a path of constant improvement of what is. Nevertheless, such an "awakening" underpins a context of harmony and peace with the world of which humans are part and with which they interact. Imbalance/harmony, good/bad, construction/destruction: all of these opposites in relation to which value judgments would usually be brought are now no more than two faces of the same thing, and feed one another.

TO REMEMBER.– Performance is synonymous with: purity, an internal action plan. This is a good re-reading of the approaches to improvement practiced in our systems. This re-reading *re-centers decisions within the perimeter of possibilities and the potential of the decision-maker*, and does so in coherence with their environment. The decision-maker does not search for the impossible or attempt to disrupt their environment. Nevertheless, although fear of the other gives way to responsibility and neutrality in this philosophy, it nevertheless requires a level of awareness that is only obtained via a *constant quest* for development and questioning.

---

9 The interested reader can consult references [NIT 01; THU 17].

To reach their ideal of purity, people in India meditate and pray, in nature, temples or monasteries. They purify themselves, fast and converse with god in the form given to him. Based on Vedism, religion has evolved. This evolution went in the direction of Brahmanism and Hinduism, which is a variant of it, then Buddhism, all cohabitating today with Islam (since the 8th century) and Christianity (from the 16th century). The thousand-year-old Vedic texts are considered to come from "God, the Supreme Being, the Absolute Personality of Godhead, is the complete person., He has complete and perfect intelligence to adjust everything by means of His different potencies total"[10]. Vedism posits the fundamental of the individual self and the universal self, which it makes exist through a "system" of gods, the Devas, who act together on the world and have mastery over it. These gods were distributed across various manifestations in the natural world and in the functions of daily life, in a framework devoid of value judgments, since each phenomenon is needed in the same way. Appearing around 600 BCE, Brahmanism and Hinduism present an evolution of Vedic practice. The Brahman, the priest, plays a central role in this, his knowledge of Vedic texts leads him to act on events by reproducing, through rituals, the action of the Devas on nature. Brahmanism recalls the existence of everything – to be an absolute or universal soul [GRI 83; HER 92] – of which the gods, humans, like any other element, are only partial incarnations. Hence, humans have not ceased to fuse with this universal soul. They seek in this sense to liberate themselves from their Karma, i.e. the values of the thoughts and acts in their lives. They die and are reborn as many times as necessary to liberate themselves from any impurity. The cycle of reincarnations stops as soon as humans have reached purity. They are then definitively delivered from their physical body and the sufferings attached to it and can fuse with the absolute.

TO REMEMBER.– The only rational thing for people to do is to *seek a better state* than the one that is.

Structuring and advocating a hierarchical vision of purity, Brahmanism introduced the notion of the caste. The respective castes of Brahmans, Kshatriyas, Vaishyas and Soudras, whose rank (level of purity) respectively decreases, each have a particular role, complementary to that of the others.

---

10 Quotation from the book *The Sri Isopanisad*: https://vaniquotes.org/wiki/The_Supreme_Being,_the_Absolute_Personality_of_Godhead,_is_the_complete_person,_He_has_complete_and_perfect_intelligence_to_adjust_everything_by_means_of_His_different_potencies.

In this case, it is the Brahmins who carry the Dharma, a set of religious, moral and cultural values for society. The portion of the population that fulfils functions outside those considered pure belongs to none of these castes. Judged to be impure, this portion of the population has no relationship with the other castes. Membership of a caste is justified by Karma. This membership is the responsibility of humans, to whom all that remains is to work to free themselves from their Karma and thus hope, in the next rebirth, to find themselves in a higher caste.

TO REMEMBER.– But how? Interiority and spirituality do not prevent an irreversible inequality in, what is more, an absolute acceptance. A compartmentalized system without a view to evolution. A system based on *the belief that each individual occupies the place they deserve at the moment in question*. A supreme decision-maker that is none other than a shared belief. Would our *beliefs* be the *basis of our decisions*?

Contemporary with Hinduism, Buddhism was born from the teachings of Siddhartha Gautama: Buddha, "the Enlightened One". Buddhism reprises the notion of the absolute from Brahmanism but liberates people from the notion of caste. Beings and things are elements of a whole, each element being reborn into another after its death, until Nirvana, which is the "state of perfect beatitude (which can be reached by contemplation and asceticism) aiming for the definitive absorption of the individual into the universal soul and the disappearance of desire" [CNR 21]. To reach this state, Buddhism posits the principle of the "middle way". This means avoiding any extreme, detaching ourselves, to achieve a sort of neutrality of the emotions and feelings, an independence of thought. Buddhism is presented as a philosophy that does not make explicit reference to the gods of Hinduism [BAN 90] and whose teaching is based on the observation of how the world and humans operate. The quotations attributed to Buddha are examples of this observation:

Change is never painful. Only resistance to change is painful.

Don't rush anything. When the time is right it'll happen.

Before you speak, let your words pass through three gates: Is it true? Is it necessary? Is it kind?

Do not dwell in the past, do not dream of the future, concentrate the mind on the present moment.

Everything that has a beginning has an ending. Make your peace with that and all will be well[11].

After its unification in the 4th century BCE, India adopted a peaceful mode of governance [REN 78]. Society was organized according to the Brahman caste system. Commerce involved the sale of its products to the Near-East, to Egypt or to the coasts of East Africa, then to Rome, with the opening up of seaways. Indeed, India was rich in agriculture and abundant artisanship. Its initial wealth came from the farming of rice and cotton, sustained by the invention of an irrigation system adapted to crops that consumed particularly large amounts of water. The farming of cotton permitted the emergence of weaving and then of a textile industry that is still strong today. In parallel, artisanship prospered through the massive production of pottery and metal objects. Driven by its culture of writing and its search for spirituality, Indian civilization provided a good number of texts. The various religious texts were written, a good number of Hindu and Buddhist stories, economic, political or mathematical treatises and in particular the invention of the figure "zero", of the decimal system (in which the numbers were constructed this time by juxtaposing digits from zero to nine) and the square root.

1.2.5. *China*

China, Confucius, Tao, the wars, the vastness… Around 2,000 BCE, the Han people acquired and developed the land between the Yellow River and the Blue River: China. Although it was regularly contested through invasions by its neighbors, the domination of the Han was sustained until the modern era, a duration of 4,000 years. This people systematically resisted division, assimilated their invaders and ensured that order was restored immediately to the territory. During the "Spring and Autumn period", which began in the 8th century BCE, China was divided into multiple states in permanent conflict. The Hans re-unified it gradually during the "Warring States period" with the formation of seven large kingdoms in the 5th century BCE, to arrive finally, in 221 BCE at a single empire. In the 13th century, the Mongols,

---

11 https://declutterthemind.com/blog/buddha-quotes/.

then in the 17th century, the Manchus temporarily dominated the territory, then the Hans took control of it again. The same was true of Europeans at the start of the 20th century, with their colonies and trading posts. It was the advent of the Peoples' Republic of China, still governed by the Han, that put an end to this latter domination in the second half of the same century.

TO REMEMBER.– Conquest, war and the law of the strongest. Just as in Ancient Egypt, a marked Chinese *identity* had allowed the system to achieve *durability*. An offensive strategy returned, with a purpose: the restoration of order. Order is to China what perfection had been to Ancient Egypt and purity to India. Performance is synonymous with *order*. The *"right" decision* is the one linked to order.

Like other primitive peoples, the Han people practiced some forms of shamanism. The shamans, endowed with higher powers, were mediators between the here and the beyond. They gave themselves the mission of being in dialogue with the environment and gaining its goodwill. This environment encompassed two worlds. The visible world was formed of nature, whereas the invisible world concerned souls, those of the – benevolent – spirits, those of demons as well as those of the dead or dying [MAT 87]. We could therefore think of a relationship between this primary shamanism and an evolution of practice and beliefs towards the cult of the ancestors, the survival of the soul and the forces of nature.

Nature and its laws structure some beliefs in China. Any perceptible manifestation by humankind is a sign of the Qi, the principle that is invisible and in perpetual movement that presides over all, which surrounds humankind. "The universe is perpetually self-creating in a constant evolution [...] from a single material, the Primal Breath [...] which is neither matter, nor spirit" [ROB 91b]. Qi is formed of the famous yin and yang, which expresses duality, and is formed in complementary fashion. A form of positioning would say that the yin unifies, integrates, uniformizes and pacifies when the yang separates, individualizes, differentiates, activates [CHE 89; JAV 89]. The relationship between yin and yang accompanies the existence of all beings as well as the development and destruction of the material world. Because of this, mutual opposition, complementarity, transformation, destruction and generation make yin and yang indivisible.

"A hundred schools of thought" were formed at the end of the Spring and Autumn period, a period of conflict for China. The aim of these schools was to find the harmony of humankind with the order of the world [GRA 15]. These schools included Taoism and Confucianism, respectively born in the 4th and 5th centuries BCE.

Initiated by the teaching of Lao Tzu, Taoism advocates for nature as a model of organization. A quest for order becomes synonymous with the quest for harmony. Harmony is based on the fact that each element of the universe has its place. A path – Dao or Tao – exists to do this. Humankind in Taoism is both at rest when they communicate with the yin, and in action when they communicate with the yang. The spirit and the heart, the intellect and the emotions, reason and instincts are balanced. Humans are neither negative nor positive, they are on the central axis, their natural position in the universe. There is therefore no need to change the course of things given the natural cycle of alteration between yin and yang: what exists today will be replaced tomorrow by its opposite [LAO 15]. This state of mind leads humans to seek to know themselves, and to find harmony themselves: "He who knows a lot about others is perhaps educated, but he who understands himself is more intelligent"[12].

TO REMEMBER.– The idea is to *find order* through a form of *mimicry of nature*. The decision-maker has the role only of *taking care of this conformity*. No effort at reflection nor evaluation, just observation and transfer, in a *complex* world, in permanent interaction and transformation.

A contemporary of Taoism, Confucianism, born out of the reflections of Confucius, advocates stability and social order. Humankind has a duty to adopt ethical behavior within the family circle as much as in the public domain. To do this, they need to raise their level of wisdom and knowledge.

> When moral knowledge has reached its final degree of perfection, the intentions are then rendered pure and sincere; when the intentions are rendered pure and sincere, the soul then enters into probity and right; when the soul has entered into probity and right, the person is then corrected and improved; when the person is corrected and improved, the family is then well-governed; when the family is well-governed, the kingdom is then well-governed; when the kingdom is well-governed, the world then rejoices in peace and good harmony. From the most

---

12 Quotation taken from the French book *Lao Tseu ou la sagesse taoïste – 125 Citations* [LAO 15].

elevated man, to the humblest and most obscure, there is an equal duty for all: to correct and to improve themselves; wherein the process of perfecting oneself is the fundamental basis of any progress and any moral development [HAS 84].

TO REMEMBER.– China offers two schools of thought on decision-making: the *Taoist decision* to act according to an *open rule* dictated by nature; and the *Confucian decision* to act according to a *canvas defined* by humankind in search of an order that they have also defined.

Buddhism arrived in China in the 1st century BCE, brought about by the opening of trade routes from neighboring India. China, where Taoism and Confucianism then dominated, adopted it just as much: "When Buddhism [...] made its appearance in China, the first translators did not know what terms to use to translate Indian concepts; they used words borrowed from Taoist philosophy. This is why, for us, Taoism and Buddhism often resemble one another" [VER 03]. For their part, those in power further appropriated Confucianism to foreground the values of work, family and country, values that were useful to them in governing.

Indeed, China is and has always been a vast, highly populated territory. Its organization, hierarchical and pyramidal, has been dominated by an aristocracy with an emperor at the top. From the lowest class, which is the peasant class, we find, in increasing hierarchical order, the artisans, the traders, the scholars (mandarins) and finally the aristocracy [MAS 96]. After the Warring States period and the return of a unified territory, the state put in place unique legal, financial and military systems, with the idea of a collective taking precedence over the individual. Until the 19th century, the territory was organized on six hierarchical levels from the province (about 20), to the sub-prefecture (several thousand). It was administered by functionaries who belonged to the educated classes, with the exception of major moments when the emperor and military took over governance [GER 97]. A system based on meritocracy made it possible to recruit functionaries with well-defined competencies and who practiced a specific written language, different from the language spoken by the people.

TO REMEMBER.– Here is a framework for *unique* and *repetitive decision-making* for a *system* with a hierarchical, calculated and precise organization, hence leaving *the decision-maker little place for free will*.

China invented and somehow still invents in many areas. It has remained the primary global economic power for thousands of years. Line culture, weeding, the first cooperatives, the improvement of labor using the metal plowshare and the implementation of hydraulic systems for growing rice were among the inventions for improving agricultural productivity. Artisanship saw the appearance of silk around 2,000 BCE and of paper at the start of the 2nd century BCE. Between the 7th and 12th centuries, China developed printing techniques and firearms. Trading these different products made exchange possible as far as the Mediterranean, with the creation of the silk route in the 2nd century BCE. In the 12th century, the abacus and compass were developed, which enabled compatibility and made it possible to guide caravans. Chinese medicine broke new ground with the first book of medicine in the world in the 6th century BCE, the identification of a significant pharmacopeia (365 remedies of animal, vegetable or mineral origin) and the codification of medical procedures (massage and acupuncture). The arts focused on painting, and ceramics and porcelains, music and calligraphy. Architecture focused both on buildings and on gardens.

Additionally, at the crossroads of Asia, to which it belongs, and the West onto which it opened less than two centuries ago, Japan has retained both these worlds. Influenced by China until the 9th century, it then isolated itself, removing any links to the outside world for a period of over a thousand years. This withdrawal led it to develop its own model of society, which was autonomous and economical, given its few natural resources. In religion, its Zen philosophy, an heir to Buddhism, pushes humans towards governance and knowledge of the self, towards respect and meditation. As far as we are concerned, the meeting of this culture with the Western world in the middle of the 19th century, then in the middle of the industrial revolution, opened the way for industrial development. In this pathway, progress is based on mastering industrial activity, using knowledge and parsimony. Thus, in the domain of merchant production, there was already a distinction between what shows value and how to develop it, and what does not represent value – waste – and how to reduce it. This desire to do better should be progressive, one improvement calling for the next and should be without interruption in a cyclical process: this was the birth of Kaizen (a continuous improvement approach). Thus, in the Toyota company, the engineer Taiichi Ohno implemented such an approach known by the acronym Toyota Production System (TPS), which can be summarized by the following recommendation: "If you are going to do TPS, you must do it all the way. You also need to change the way you think. You need to change the way you look at things" [OHN 88].

TO REMEMBER.– A *methodological framework* for making decisions based on the *belief* that the success of change means proceeding *step by step* and *continuously*. One condition: mastering the system subject to improvement. One means: continuous learning from success and failures. One method: cyclical and permanent. *Little place for free arbitration by the decision-maker*.

To finish our discussion on this civilization, let us take up the words of the Chinese master:

> Many young westerners interested in Zen have said to me; it is a form of Zen emanating from Japan, but, you know, it is a pure Chinese creation: we call it chan. Its main principle is Taoist: you have to discard your thoughts and your beliefs… The aim is to be aware of a thing without being aware of it; to create the void within yourself. To arrive at this, Buddhists and Taoists have developed methods based on breathing. This is why, each morning, before we set off, I ask you to remain sitting for a moment, unmoving, before the countryside. This position requires long training but the day you succeed it in fact becomes an illumination. You still won't live differently to others, but you will gain another perspective of what surrounds you. The unaware will then speak to your awareness [VER 03].

1.2.6. *Ancient Greece*

Greece, its democracy, its science, its philosophers… What does it mean to be Greek? The answer to this question – an identity – is the keystone of Greek civilization.

Ancient Greece was born at the start of the 2nd millennium BCE, in a context of migrations between Anatolia and central Europe. The first cities emerged in the 17th century BCE, providing a city setting for this intermingling of peoples. A turning point took place at the end of the 15th century BCE, with the conquest of the Dorians, a warrior people with superior weapons. Over a century, the cities and countryside were ransacked, leading to emigration towards Asia Minor. There then came a period of re-organization lasting seven centuries, of which a hundred cities were the result. At the same time as they unified to combat external threats, each city sought to secure its access to resources. This was done through an attempt to extend sovereignty based on fighting with neighboring cities and the installation of colonies along the Mediterranean basin. In the 6th century BCE, four cities dominated and structured Ancient Greece: Athens, Sparta, Corinth and Thebes. While still being rivals, because of the scarcity of resources, they conferred on themselves, in a quest for hegemony lasting two centuries, supremacy over the whole of the Mediterranean. But constant internal competition eventually generated weakness and decline. In fact, in the 8th century BCE, a thousand kilometers from Athens, Etruscan tribes unified to found Rome, in a region then fragmented and occupied by numerous peoples: the future Italy. The development of Rome was linked to a powerful army, the legion, whose watchword was "conquest". Relatively near to Ancient Greece and its colonies, Rome took them as a model and conquered them in 146 BCE, while "becoming" Greek[13]. Indeed, "Captive Greece took captive her savage conqueror and brought the arts to rustic Latium"[14]. This quotation from Horace attests that, although militarily and politically dominant, Rome took on the attributes of ancient Greece, the model of which extended, beyond its own borders, to an immense territory, covering the Mediterranean basin as well as other northern territories.

TO REMEMBER.– *Performance* had been a question of criteria and viewpoint. Defeat at one level meant re-naming at another. The *range* of actions and so of *decisions* could be unpredictable and multiple. In addition, *the non-robust quality* of a system did not go hand in hand with its performance. A *decision that looks outside* the system, neglecting any form of interiority that could have allowed steps towards internal improvement to be implemented.

Æolians, Dorians, Ionians, etc., the many peoples that co-existed in Greece each founded cities that, with the successive occurrence of exchange, internal wars and migrations, encountered one another, mixed with one another, shared a language and culture, but still remained rivals. Greece was a fragmented whole, sometimes a whole facing an external environment,

---

13 Within the framework of this idea, contemporary Europe, the distant successor of ancient Greece and Rome, speaks of the Roman-Greek world "that belongs to the Greeks and to the Romans" [CNR 21].
14 https://www.goodreads.com/quotes/6598477-captive-greece-took-captive-her-savage-conquerer-and-brought-the.

sometimes fragmented in internal rivalry[15]. From this community, formed of a destiny of internal confrontation and victory against the "Barbarians", i.e. non-Greeks, an identity was born. A feeling of superiority succeeded this identity, first military and then political and cultural. It was this feeling that conferred on Greeks complete freedom to govern their decisions.

TO REMEMBER.– Here is a decision model involving *multiple decision-makers free in their decisions,* on the one hand, and, on the other hand, involving the defense both of a *global* interest and of *individual interests from negative interactions*.

"Faced with the uncertainties of existence, individuals and communities turned to the gods" [PIR 18]. In a few words, the scene of the beliefs of the Greek world is set. The gods were thought to be in continuity with humans and carried a Greek identity based on superiority, exploits and rivalry. The gods and also the heroes, at the intermediate level, often born of the union of the gods and men, exalted a character of power, courage, intelligence, respect and loyalty as well as domination and authority. Mythology did nothing other, relating the adventures of the gods, the heroes or remarkable men, than offer a complete narrative in the guise of a reference for Greeks in leading their lives. Greeks could grasp opportunities and avoid threats, their strategy of seeking friendship being contrasted with the strategy of pursuing the rivalry of others: a precursor vision of future game theory, implemented to manage conflicts in a co-operative environment [NAS 51].

TO REMEMBER.– A cult of *power* and the search for *exploits* were the purposes of the system.

Beyond these sentiments of power and rivalry, the Greek people were strongly conditioned to have what we today call a "logical" mindset. Indeed, Ancient Greece believed in the concept of *logos*, i.e. in the law based on the coherence of all things in the universe. "Logos is what links phenomena to one another, as a phenomenon of ONE universe and what links discourse to the phenomenon; logos is a link" [AXE 62]. The Greeks' vision of their

---

15 Plato stated, "We shall then say that Greeks fight and wage war with barbarians, and barbarians with Greeks, and are enemies by nature, and that war is the fit name for this enmity and hatred. Greeks, however, we shall say, are still by nature the friends of Greeks when they act in this way, but that Greece is sick in that case and divided by faction, and faction is the name we must give to that enmity…" https://digressionsnimpressions.typepad.com/digressionsimpressions/2018/09/on-plato-and-kant-and-war-conduct.html.

world was thus colored by a need not only to frame, explain and deduce each thing from another (as in the ancient Near-East of Ancient Egypt), but also to "express" these things as well as their links (associated them with names). The term *logos* would also be translated a little later by the Romans by the term *ratio* which in Latin means that which appeals to sense, to reason.

TO REMEMBER.– A "rational", "logical" and "explicit" framework encompassed the parameters of all decision-making.

The first Greek cities were governed from the "palace" where the king nominated functionaries in charge of administrating the territory. After the Dorian period, the Greeks governed in diverse forms such as the aristocracy, where power was allocated according to birth, the oligarchy, where power belonged to the strongest and richest class, or tyranny, which saw one individual take absolute power alone. And it was in the period of the four dominant cities that a new form of power saw the light of day in Athens: democracy, where the rights of all citizens were equal. This was the advent of politics.

TO REMEMBER.– *Who decides*? The noblest? The strongest? The richest? Everyone? As many choices as decision-makers, which is still the case today.

Fishing, agricultural and artisanal production gave rise to maritime trade through the mastery of ship building and navigation. The development of colonies amplified this exchange of foodstuffs, materials, pottery and textiles. This wealth allowed artistic development that began with the construction and decoration of the palaces of the first cities and then extended to ceramics and sculptures, temples and city monuments. This aesthetic desire and the search for perfect proportions spread to the worlds of science, poetry, theater and philosophy: it was the latter that Pythagoras first mentioned, defining himself as a "philosopher, not one who claims to possess wisdom, but a man who strives towards it". This world thus presented two faces. One of the faces, that of the governors and warriors, conveyed an idea of power and superiority. The other face, that of the sages, was full of humility before the mysteries of humans and of the world. This was echoed in the writing of Socrates: "The only thing I know is that I know nothing, and I am not quite sure that I know that," and know thyself[16].

---

16 https://www.goodreads.com/quotes/1176881-the-only-thing-i-know-is-that-i-know-nothing.

In this search for wisdom and in conformity with the *logos* vision of the world, ancient Greece introduced the forerunner of what is now called the scientific approach. The Greeks advocated an intellectual approach, the method of which was developed by Aristotle[17]. "Logic", in its principles of deduction and induction, was adopted. An approach was called scientific if it had three respective stages: observation, theorization (modeling according to a theory) and then experimentation/validation (in reality). For example, we cite Thales who laid the foundations of geometry by observing the Egyptian pyramids; or indeed Eratosthenes who founded geography by measuring the shadow to deduce the circumference of the Earth.

We emphasize the Greek model of decision because Western practice in this area owes much to it. Athens invented democracy after a financial and social crisis rocked the city. A series of reforms was embarked upon, calling for the consent of the citizens. This led to a constitution that described the role of political institutions and the method of choosing their members. In this sense, Socrates decreed that wisdom required the governors of the city to deliberate before making a decision, in other words to weigh up different viewpoints and make a collective decision. This construction should then, in conformity with the thinking of Aristotle, be made under the constraint that it conforms to morality. The decisions made should then result "from a positive orientation of reasonable wishes (*boulesis*) of the agent towards good. Reason and desire appear, from this point of view, to be inextricably linked" [MOR 17].

TO REMEMBER.– *Deliberation*, *process* and *method* were associated with making decisions.

1.2.7. *The Arab–Muslim world*

The Arab–Muslim world, Islam, poetry, the magic of the Orient... Here was a civilization based on the reach of its religion. Indeed, in the 6th century, the Arab peninsula, shared between merchant people and tribes of nomadic herders – the Arabs[18] – was subject to influences from Rome, Persia and the Near-East. In this soil where Judaism, Christianity and polytheism co-existed, in 611 a man, Mohammed, unified the tribes in the

---

17 From the Greek *methodos*, pursuit, (re)search, formed of *meta*, towards, and *hodos*, path.
18 In Yemeni, the word *arab* means nomadic herder [ROB 91a].

name of a new religion and its principles, Islam. No territory was associated with this new religion but instead to a notion of community (*ummah*) to which all Muslims belonged[19]. After the death of the prophet Mohammed in 632, his successors (*khoulafas*) spread, in a few decades, as far as Damas, the territory, called the caliphate, from India to the Atlantic Ocean. From 750, the Abbasids made Baghdad their capital and extended over the whole territory for five centuries. In the 19th century, the Ottomans gradually took control over the greater part of the territory and added Southern Europe to it, to reach its greatest extent in the 18th century. The 20th century saw the decline of the Ottomans and the fragmentation of the Arab–Muslim world into a number of independent territories. Beyond an organization in the form of a caliphate, empire or independent countries, the Arab–Muslim character remains essential for the whole community today, whether in multi-faith Lebanon or other countries in Asia or continental Africa that claim an Arab culture or Islam.

TO REMEMBER.– A *system* whose construction is based on a *belief*, and whose *borders* are delineated by the practice of the latter.

Humans in the pre-Islamic Arab peninsula were shaped by their desert environment. The family and the tribe were the focal point of a society formed by laws of solidarity and hospitality, competition and pillage. Each tribe believed in their own gods, which had to be venerated to obtain help and avoid wrath. Islam was to modify this society not only in its beliefs but also in its recourse to the precepts of religion to judge what was permitted, the "good", and what was forbidden, the "bad". Nonetheless, the believer was left alone to interpret and decide their choices. This quotation from Ibn Rochd (Averroes) is emblematic of this mindset:

> Since therefore this religion (sharia) is the truth, and calls for the practice of rational examination that ensures the knowledge of the truth, then we, Muslims, know with certain knowledge that demonstrative examination does not lead to any contradiction with what the religion says: since the truth cannot be against the truth, but agrees with it and bears witness in its favor.

Humans become responsible for their actions, for which they must answer, since their conscience is complementary to their faith.

---

19 Which means to submit to the rules and laws of the Muslim religion, repeated in its holy text (the Koran) as well as in the rules of its prophet, the Sunna.

TO REMEMBER.– A *bipolar* and *absolute* vision of performance. To satisfy oneself with the best and to fear the worst. The making of decisions based on a set of beliefs, values and rules. A model of decisions *devoid of any hierarchy or centralization*. A permanent learning of rules being adopted.

Arab–Muslim people were also eastern and carried with them an aspect of joy, of the sacredness of life, and at the same time of renouncement and fatalism. Because of this, the Koran teaches, "Work for your worldly life as if you are living forever, and work for your Hereafter as if you are dying tomorrow"[20]. Poetry could magnify the Arabic language by exalting, according to its very precise canons, the beauty of the Orient, love, friendship or feasting. The decoration and the ornament came to sublimate any work to tint it with art. Ibn Rochd applied this vision to medicine, then considered an art, for which he advocated a deductive logic:

> The art of medicine is an art that acts based on true principles; we seek through it to preserve the health of the human body and to eliminate disease as much as possible in each body. The purpose of this art is not inevitably to achieve healing, but rather to act appropriately in the appropriate time and extent, while waiting to achieve its purpose, as happens in the art of navigation and the command of the army[21].

TO REMEMBER.– *Intensity* in the making of decisions and also *detachment*.

From its expansion, the Arab–Muslim world was divided into provinces, the administration handling taxes, treasury, estates, justice, arms and letters. By its nature, this world was very diverse and endowed with many languages. However, the Arabic language was the one used in administration and in the practice of religion. In the Ottoman empire, administration was centralized and governed by a system of meritocracy.

For problems whose solutions could not be found in religious and reference texts, Arab–Muslim people developed practices for seeking consensus (*ijma'*) and reasoning by analogy (*qiyas*). Having affirmed that "the whole Koran is nothing but a call for examination and reflection, an awakening to methods of examination", Ibn Rochd subscribed to this

---

20 https://www.islamweb.net/en/fatwa/91760/work-for-your-worldly-life-as-if-you-are-living-forever.
21 https://philosmus.org/en/archives/1901.

logic and knew many detractors who did not agree with his vision of Islam. Today, so-called "Islamic law" continues to inspire many countries in developing their constitutions or even in treating more general legal cases.

TO REMEMBER.– A *decision model* based on *reflection* and *comparison* for seeking *consensus*. A constantly expanding system of rules.

The economy of the very early Arab–Muslim world relied on animal husbandry, agriculture and commerce. Husbandry, sheep and goats were carried out by the population's nomads. Agriculture was ensured by those who were settled, and business by its merchants. Artisanship developed and then became specialized in the different provinces. In the domain of knowledge, the Arab–Muslim civilization found and translated forgotten texts by Greek philosophers and conveyed discoveries emanating from Asia to the West. In the domain of science, advances combined respect for the precepts of the sacred and the quest for truth. These advances progressed and were spread throughout the region and beyond. The phrase, "the Arab miracle" is used to describe the zenith of this civilization, at the forefront of mathematics, optics, mechanics, chemistry, astronomy, geography, history, philosophy, poetry and medicine.

1.2.8. *The Western world*

To what extent did the Renaissance shape how the Western world exists today?

When the Middle Ages ended, the Western world was Christian. Humans belonged to a community of believers marked by the values of good and bad. Notions of sin and guilt for faults committed were related by the Church and its clergy, who guided the population, as the shepherd guides his flock. This framework crumbled in the Renaissance, during which this world discovered its Greek roots, exchanged with the Arab–Muslim world and projected its horizons as far as Asia. It is this (re)naissance (re-birth) of a new world that saw, lived and thought more widely that we will now describe.

Let us make a slight change to the description adopted until now to describe this Western world by means of a tool ordinarily applied to improving quality in businesses: the "five Ws" method[22]. *Who?* The European States whose political forms evolved from authoritarian monarchies, until the 19th century, to democratic systems. *What?* An essentially economic and intellectual activity. *Where?* On a territory first circumscribed to Europe and then extended to its colonies. *When?* From the Renaissance, at the end of the 15th century, until today. *How?* Through ideas, commerce and weaponry. *Why?* A search for the liberation of medieval practices, in a religious context instilled by Christianity.

What should be noted in this circumscribed world, as described above? Two aspects appear to us, essentially: its materialist vision of the world, on the one hand, and its intellectual movement, on the other hand. The quest for material wealth was in fact at the root of colonialism as well as of intensive exploitation of the natural world. As for the intellectual approach, it would lead to a scientific and philosophical understanding of the world. Although this understanding was intended to liberate humans, over the centuries it was used partly for humans' material expansion.

Materialism, which affirmed the primacy of matter over the mind [LAR 21], accompanied Western people. A consequence of their past, it advocated possession as a means of existing and in fact as an end in itself. This possession, advocated by capitalism since the 16th century, took both individual and collective forms [BRA 81]. In fact, some personalities, formerly the nobles and middle classes, today businesspeople or those skilled in the virtual economy, transformed and still transform this quest into a search to maximize this individual wealth. They thus came and went against the vision of one of the forerunners of economic thought, Smith, who saw in the individual's initiative only a contribution to the general interest [SMI 76]. Based essentially on business, then on industry, their possession also chimed with colonization, the extension of territories, the subjugation of populations and of natural resources. This paradigm of domination created the power of Western people, they met with their limitations only at the start of the 21st century.

---

22 The "five Ws" method (who, what, where, when, how, why) is used in business in the domain of quality. In principle, it consists of taking the time to define a problem before proceeding to solve it [CHA 06].

This way of doing things, which has directed the Western world from its beginnings, was never unanimous. A number of opponents – from the protestants of the reformed Church, appearing in the 16th century to the alter-globalists of today, via the revolutionaries of the 19th century – defended a society based on values such as the collective interest, openness, the equality of peoples, poetry and the arts. Indeed, in a more basic fashion, the primacy of matter over the mind was called into question by the doctrine of humanism, which places "humans and their condition at the heart of its concerns. An intellectual movement that developed in Europe during the Renaissance which, with a keen, critical appetite for knowledge, was aimed at the fulfilment of humans rendered more human by culture" [CNR 21]. This break with the Middle Ages led Western humans to favor an understanding of the world through their intellects. Humans were no longer content to feel and accept the world that surrounded them; they needed to understand it before they could act. We cite, for example, Nicolaus Copernicus, physician, administrator, military leader and astronomer, who dared to think of the heliocentric model of the solar system in the 1510s. Thus, although materialism is the alpha of this Western world, humans who think, who create, who investigate the domains of sciences, culture and art, are its omega.

This intellectual movement (still present today) is deployed in all domains of human activity. It takes all forms linked to analysis, thought and creativity. It involves knowing, understanding, thinking and debating, defining, modeling, observing, asking questions and solving problems, etc., as well as turning to the muses for artistic or poetic expression. In this respect, this movement takes up the thread of Greek thought, interrupted for a millennium, and enriches it with multiple forms of knowledge, from India and China as well as the nearby Arab–Muslim world. Humans, free to think and to create, opened up to all fields. Examples such as that of Leonardo da Vinci, both a genius painter and a visionary engineer, or William Shakespeare whose work transcended epochs and countries, or even Albert Einstein, whose perfection of thought erected no barrier between sciences and ethics, illustrate this new way of thinking.

A determining stage of this movement was the era of the Lumières illustrated by Emmanuel Kant in the following way:

> What is the Enlightenment? Man's leaving of his minority, a departure for which he is himself responsible. A minority, i.e.

the inability to use his understanding without the direction of others, a minority for which he is himself responsible since its cause lies not in a defect of the understanding but in a lack of decisiveness and courage to help himself without the direction of others. *Sapere aude*! Have the courage to help yourself by your own understanding. Hence the Lumières motto.

Why accord interest to the Lumières in our quest for a decision? Because it is to the Lumières that we owe the methods as practiced in the Western world today. It is to the Lumières that is owed above all an approach centered on reason in the prolonging of *logos*, the Ancient Greeks' concept of logic translated into Latin by the term *ratio*. This can be stated without any inaccuracy, considering that it was the Lumières who incarnated the intellectual approach of the West.

In fact, to construct their own knowledge of the world with the help of the decision protocol mentioned previously, humans rely on three principles. The first principle stipulates that access to knowledge is achieved by reason (which includes intuition, evidence, deduction, the senses, the imagination and memory). The second principle posits that it is the very essence of humans to think for themselves, with this thought seen as the use of reason, seen as the very proof of their existence. Finally, the third principle concerns the uselessness of recourse to faith to obtain knowledge.

Fidelity to these principles required of humans a systematic respect for the stages to be followed to "reason well". In this respect, it prolonged the practices of the civilizations of the ancient Near-East, Ancient Egypt, Ancient Greece or of the Arab–Muslim world, which all sought to understand the world according to their own reasoning. In this sense, the philosopher and mathematician René Descartes proposed a universal scientific approach: Cartesianism. This approach sought to reconcile human desire with the reality of the world perceived through accessible knowledge, in a continuation of the *boulesis of* Aristotle, who knew that desire and reason were indissociable (see section 1.2.6). So, Cartesianism became "the" method "for properly guiding one's reason and seeking truth in the sciences" [DES 37]. In this sense, humans argued with themselves, for they had known since Socrates not only that they do not know everything, but that they know exactly what they do not know, and that they consequently decide with an imperfect knowledge of the world (see section 1.2.6). This is why, according to Descartes, humans are incited to follow a protocol to "properly conduct reason". Thus, Descartes defines "method" as a set of reliable rules which are easy to apply, and such that if we follow them exactly, we will never take what is false to be true or fruitlessly expend our mental efforts, but will gradually and constantly increase our knowledge until we arrive at a true understanding of everything within our capacity. "As for the method, I understand by that, particular and easy rules whose exact observation will mean that no-one will ever take the false for the true, and that, without uselessly expending any effort of the intelligence, he will arrive, by a gradual and continual increase of knowledge, at true knowledge of everything he is capable of knowing"[23]. Being Cartesian is explained today by a method of being that "presents rational, rigorous and methodical characters proper to the intellectual and spiritual approach of Descartes" [CNR 21]. In other words, whoever applies Cartesian principles in their reasoning becomes Cartesian, to the point that the word has become almost synonymous with the scientific approach.

TO REMEMBER.– Beyond the methodological principles that characterize his work, Descartes thus revealed to us all, three centuries before Japanese engineers, that to be effective, it is enough to be efficient and to subscribe to a continuous improvement approach.

Concerning decisions, the naive vision presented in Figure P.8 conforms to this method, in which the "Before" stage corresponds to observation, the "Decision" stage to the distinction of true from false according to a given protocol and the "After" stage to the production of new knowledge. This continuity, this temporality and this progressiveness in the different stages of the act of deciding thus explain why the term "decision process" can be identified in it. Three centuries later, Simon proposed nothing other in a decision model, the reasoning of which was based on respect for a procedure [SIM 55].

TO REMEMBER.– Decisions are defined as the result of a *process* based on *reason*. The input being identified as *knowledge* of the object of the decision and the *stages* being linked to *debate* and to *action*.

---

23 https://plato.stanford.edu/entries/descartes-method/.

The West does not differ from the evidence provided by these civilizations, which showed that decisions have always been in the nature of humans. The aspect of originality led by the Lumières lies in a practice of reasoned decision resting on scientific methods and tools. Thus, the notion of utility proposed by Bernoulli aiming to retain the solution in the hope of maximum gain, or the modes of scrutiny for electing the candidate preferred to all the others of Nicolas de Condorcet, aims to systematize the result of the decision process. The idea is therefore that this result is independent of the decision-maker, in other words that two different decision-makers would make the same decision. More recently, new decision paradigms such as those involving multiple decision-makers with their own objectives, consideration of risks in the decision, the consideration of uncertainty or the difficulty of describing the problem completely have made it possible to enrich these methods and to develop new ones. These innovations have led in particular to questioning the idea of an independent decision by the decision-maker. But whatever the case, it is already clear for this current of thought that the decision needs to be analyzed, according to some, while others say it should be aided by methods among which, perhaps as a side note to the reasoning, the decision-maker will have to choose.

TO REMEMBER.– Beyond having provided the frameworks necessary for making decisions, the process and the methods, the Western world insisted on both notions of decision and decision-maker, the decision-maker being in a situation of responsibility (see the quotation from Kant). It is therefore really a question of the essential decision-maker/decision pairing.

1.3. Conclusion

In the guise of a summary, Figure 1.3 reprises the civilizations described throughout this chapter (Figure 1.2) and suggests for each of them, with regard to the notion of decisions, the most notable facts.

Thus, unsurprisingly, humans have sought, whatever the space and time, to satisfy their objectives and to control their fears at the same time. Impacted each time by the context, the expression of these objectives and these fears take various forms which are sometimes extreme. And each time, the decision was a sort of search for compromise between these two strong aspects of human nature.

Although some civilizations have become extinct or been transformed, others still survive in the modern era. Moreover, depending on the impression made by each, the decision-maker in us will recognize themselves in this or that civilization and in the intentions it conveys. We will have seen therefore that the context was impactful on the act of deciding (Figure P.7), through both the decision-maker, their objectives and ultimately their decisions. But finally, the same invariants still characterize the decision. "Hierarchized, centralized, formalized", such is the decision that humans can make, both to reassure themselves and to respond better to their intentions. The hierarchy foregrounds aspects linked to decision-makers' choices and to the links between them. The centralization is fairly corollary to this hierarchical vision, reinforcing this idea of being able to center the decision on a single decision-maker, the "chief decision-maker", who will assume all the glory along with all the risks. The formalization responds, for its part, to considerations such as the need to understand, the need not to be fooled, to do our best, etc. In this respect, process and methods become the naturally indispensable tools for deciding. They would in some ways dispossess humans of their ability to "decide well" but would at the same time guide them when indecision and non-decision (see Preface) would knock at their door. Other ways would have been possible, naturally, for exploring this notion of decision, i.e. its link to performance and to action, notions particularly close to and important to decisions. Nevertheless, for reasons of cohesion, this way will be set aside for the present moment and will only be considered indirectly in the remainder of the book.

Processes and methods will be imbued with the spirit of centralization and hierarchization that humans have given to decision and are the subject of the following chapters. Honoring this process, a general framework is addressed in the next chapter.

Decision and Decision Context 35

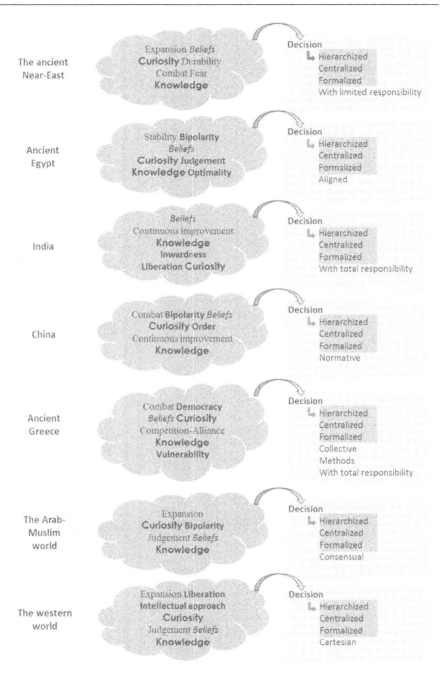

Figure 1.3. *Civilizations and notable facts. For a color version of this figure, see www.iste.co.uk/berrah/decision.zip*

2

Decisions: The Process

How do we decide what to do?

H. Simon

2.1. Introduction

One of the many characteristics of the notion of decisions is that it results from the implementation of a process called the decision process[1]. Aristotle and Descartes after him had in fact stated this well before we did (see section 1.2.8).

We distinguished in our previous discussion the act of deciding (decision-making) from the decision, strictly speaking.

Decision-making thus illustrates the action, and the decision illustrates the result of this action. Decision-making can therefore be identified with the process describing the mechanism by which a decision-maker's intention evolves, for a considered system, towards a decision that they make for this same system. Among the key stages of this mechanism is the stage of choosing the actions to carry out, in coherence with the objectives stated during the previous stages. The objective reflects the concretization of the intention. And the decision becomes the preliminary artifact to carrying out

---

For a color version of all the figures in this chapter, see www.iste.co.uk/berrah/decision.zip.

1 For the sake of clarity, we will use the expressions "decision process" and "decision-making process" without distinction in this book.

the action chosen by the decision-maker for the system, in view of reaching the fixed objective. The decision is the process concentrated in one image (Figure 2.1).

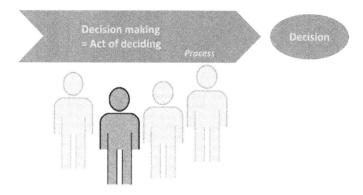

Figure 2.1. *Decision-making and decision*

The decision is the expression of a choice in view of an objective. The process is the pathway that makes it possible to arrive at this choice. In the Foreword of this book, we addressed the diversity of such a process given the different facets of the decision (Figure P.7). Described previously in an intuitive and general fashion (Figure P.8), such a process aligns with a universal practice associated with the principle of having an intention before deciding, then acting after a decision is made (see Preface). The decision process thus connects, respectively:

– a preparatory stage to the decision that identifies the system (the subject of the decision) and the objective;

– a decision stage that depends on how the decision is made;

– a stage that addresses the implementation of this decision and the expression of the associated performance, this expression is obtained on the basis of a comparison of the results achieved with those expected.

Figures 2.2 and 2.3 respectively represent and summarize this sequence precisely. This is how the decision-maker addresses any decision to be made. Naturally, questioning the chosen objective or approach will initiate new processes.

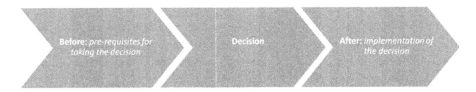

Figure 2.2. *A naive vision of the decision process*

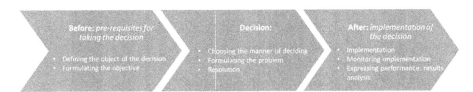

Figure 2.3. *A clear vision of the decision-making process*

This is, for now, the essential part of our reflections on the decision process. But what exactly is this process that allows us to move from imagination to reality? Without allowing us to arrive at a definition, the contextual analysis carried out in Chapter 1 nonetheless confirmed the "role" of the decision-maker. More particularly, that of their roles and reference practices in the way this process evolves. These latter are dictated by their environment and the values of their civilization. One decision-maker will thus consider it vital to seek the "best possible" (this was the case for Ancient Egypt (see section 1.2.3)) or the "best that can be done" (this is the case in India (see section 1.2.4) or in China (see section 1.2.5)). Another decision-maker will base this process on a search for all the elements necessary for a deliberation (this was the case in Ancient Greece (see section 1.2.6)). The "for" and "against" are therefore studied in detail in a purely intellectual approach in what has already been called *logos* or *ratio* (see section 1.2.6). This is moreover how the Western world defines decisions, as the result of a process based on reason. It associates the term rationality to indicate the scientific nature of the mechanism adopted for decision-making (see section 1.2.8). We will see that these nuances in intentions impact the understanding of the three blocks of the decision process.

The discussion in this chapter is therefore an analysis of the decision process. After making some general remarks on the concepts associated with the notion of process (section 2.2), we will recall, first, the essential ideas relating to this notion (section 2.3), in order to position the sequence of three blocks introduced previously. Second, the visions proposed in the literature for the decision process are explained (section 2.4), distinguishing essentially an optimization approach and a compromise approach. We will return in both cases to the three blocks introduced to make a comparison between what we believe is required in each block and the response from the state of the art (sections 2.5 and 2.6). There will be a focus on the decision-maker's involvement, and equally a parallel with the industrial world and its approaches to making decisions will be proposed using some illustrative examples. Some lessons for the remainder of the study will then conclude this reflection (section 2.7).

2.2. Why a decision process?

To begin, let us return to the quotation from Renard with which we began our reflections: "Once my decision is made, I hesitate for a long time". Couldn't this surprising way of making our decision then questioning ourselves be read differently? If Renard had mentioned his hesitation, this could mean that the decision made – hence his choice – did not perhaps satisfy him entirely, or that the impact of his decision (i.e. of his choice) was not known to him or was not entirely positive. Explaining the decision-making process will be even more significant in situations in which the decision-maker seeks to form their decision: when they are not "too sure" of it, when they are not entirely in control of their system, when the fixed objective arises from a deficiency or a need that must be met, when this objective may be reached to the detriment of other parameters, etc. These are all considerations that would prevent this decision from being made immediately; it may present "fors" and "againsts", etc. By implementing different stages of a process to make a decision, we could adopt a "rational" approach and guard against the fear of "making a fool of oneself", of "cutting corners", of "making the wrong choice" or of "making the wrong decision". It is thus on the "quality" of implementation of this process that the "quality" of the decision made will depend.

Let us continue with our interpretation of Renard's attitude. As a decision-maker, he therefore notes a long hesitation *after* making a decision. This hesitation could mark a break between the decision-making block and the decision implementation block. It might also suggest to us, in this implementation, a preliminary stage… of uncertainty? Of risk analysis? However, Renard probably put himself, more or less consciously, in a frame of mind that could really allow him to picture the consequences of this decision. There may thus be some decision-makers who act in some ways like chess players, feeling a need to dub their pieces, to move them but without letting them go, to better evaluate the benefit of their moves. They will hesitate, then confirm their choice by placing the piece or returning to their initial position before placing it definitively. This attitude, which may be evidence of uncertainty, of fear, etc. illustrates a kind of gap between the decision-maker's intention and/or belief and their decision. In this sense, we might expect from the decision process a search for this alignment between the initial intentions and the final decisions. And, in any case, a "de-complexification" of the act of choosing which, *ultimately*, engenders all these questions.

Moreover, the definitions of the term decision say nothing more or less than that this form of reflection conditions the choice. In fact, we find this idea too in both propositions from CNRTL, mentioned below:

– "*action de décider quelque chose ou de se décider, après délibération individuelle ou collective*" (the act of deciding something or of making a decision, after individual or joint deliberation);

– "*choix réfléchi de l'une des issues au terme d'une délibération*" (a reflected choice of one outcome after deliberation).

In other words, to decide it is necessary to reflect, to propose, to deliberate, as mentioned previously, "to reflect, to discuss, to organize and, *ultimately*, to choose" (see Preface). Naturally, this approach to decisions belongs to a typology of decisions and a particular vision of the notion of decisions. The analyses carried out in Chapter 1 have in fact allowed us to consider the impact of cultures as well as the weight of civilizations on the ways in which decisions are made. The approach foregrounded above correlates well with the (Ancient) Greek vision in which decision-making had been associated with a deliberation mechanism (see section 1.2.6). It is also correlated with the Western vision, in some ways inherited from this same Greek vision, which for its part links decisions to an intellectual approach carried out by the decision-maker and based on their thought, reflection and logic (see section 1.2.8), the rationale adopted guaranteeing the correctness of the decision made.

We note moreover that the first definition relating to the decision emphasizes a dynamic aspect that is linked to the process and which identifies what we consider to be decision-making. For its part, the second definition is linked rather to a characterization of the decision that results from this process. These visions unsurprisingly strengthen our distinction, mentioned before, between decision-making (making the decision) and the decision itself (Figure 2.1). They introduce, beyond this, the dynamic notion of process, the result of which is this static notion of decision.

Even though it remains little addressed as such, the notion of the decision process is at the center of many studies, since this process in fact presents the benefit of federating all the stages that allow a decision to be reached. And although in some cases the benefit of this federation may be small, it will be quite otherwise in other cases. Many ways of carrying out the decision process have thus been proposed in the literature [ROM 01]. They go from intuition, which points towards a solution without an explicit mechanism for reflection, to supporters of an exclusively intellectual approach, who propose setting the decision in an "equation" of the decision in a rational vision of the result obtained. The process therefore amounts to transcribing the "Before" stage of the decision into formal language, then processing it formally to obtain the "Decision", thus leaving no place for a "Hesitation" stage. Inviting us to take another perspective, the observations made in the 20th century by Simon and his successors on "effective" decisions in organizations have led to a reference formalization of the decision process in a rational vision, this time of the process followed. Thus the observations made suggested the resulting formalization pointed towards a hypothesis that creating a mathematical equation of some situations was impossible. In conformity with the contextual analysis of Chapter 1, it may be interesting to underline that such a search for formalization can still be evidence of a desire for standardization (Figure P.7) and so for control by humans of their actions to reach the best choices.

Within this logic, it will no longer be surprising to state that there – situations, businesses, societies – where decision-making provokes reflection, choice and eventually debate, methods and tools have been developed as decision aids. Some have become, as we will see in the following chapters, dedicated to making decisions, whereas others are only indirectly linked to them, presenting processes similar to those of decisions. This happens to be the case with management and continuous improvement tools [DEM 82; IMA 92] generally dedicated to the context of production.

Choice and reflection thus seem to be the two keywords that characterize decisions. Process is another.

2.3. The notion of process

A number of application fields have recourse to the notion of process to model, plan, order and govern their activities, whatever they may be. The notion of process is fairly intuitive and evokes a shared understanding of continuity, time and the progression of stages. The term process draws quite rightly on its Latin root *procedere* (to advance) and generally conveys the semantics of dynamics, sequencing and progression. More precisely, the CNRTL defines the notion of process as follows: "All the successive operations, organized with a view to a set result." As for the *Grand dictionnaire terminologique*, it gives the following definition: "All the logically interlinked activities that produce a set result". Still beyond the decision process, the ISO 9001 standard proposes for the world of business and industrial organizations to see a process that conforms with the following definition: "A *process* is a set of activities that are interrelated or that interact with one another. Processes use resources to transform inputs into outputs[2]" [ISO 01].

Finally, all of these definitions converge into some simple ideas:

– a process is formed of operations or activities;

– these operations are linked to one another by a temporal link (synchronicity) and/or by a logical link (cause–effect link);

– the process transforms the input elements into output elements;

– the outcome of the process should correspond to a set result.

The intuitive vision of the Foreword, which spoke of "the transformation of an idea, a feeling, an objective, an observation, into a decision" corresponds entirely to the definition of a process. Its output element is the decision when its input elements are the idea, the feeling, the objective and the observation. As for the activities, they will correspond to the different blocks in Figures 2.2 and 2.3, in a logic of progression and succession. These blocks/activities will need to be detailed before they can be operational.

---

2 https://www.praxiom.com/process-approach.htm.

Detailing these activities means attempting to provide the decision-maker with a sort of guide to answering the question: How do they make this decision? We will see the answers provided by the state of the art in this. And this cannot be done without evoking this notion of rationality, systematically associated with the decision-making mechanism in the Western world, a referent in this book.

2.4. Decision-making: rationality or intuition

Beyond the fact that it can simultaneously incarnate the "before", the "during" and the "after" of the decision, the decision process in some way identifies the way in which the decision will be made. This way of "going about it" affects the "nature" of the decision that results from it and vice versa (see Preface). In conformity with the usual practice of manager engineers, we choose in this reflection to place decision-making within the framework of logic and reason offered by Descartes' school of thought (see section 1.2.8) and that of Aristotle before him (see section 1.2.6). Because of this, we will consider the definition of the decision process from this perspective. More precisely, in a continuum of this approach, it is the concept of rationality mentioned previously that will embody this vision. In conformity with CNRTL, rationality presents the "character of what is rational, logical". It comes in some way to "signify" the reasoning inserted into the decision process. It will be set, as appropriate, based on adjectives such as total or limited, substantive or procedural. This would then be called a decision process for a rational decision. A rational decision is a decision "made using reason", as opposed to a decision called an intuitive decision, "made in emotion" [LAZ 91; DAM 94].

The rational decision is defined without ambiguity and implies:

– an approach structured in the prior definition of a problem/a question [TSO 06];

– the existence of one (or more) solution(s)/answer(s) to this problem (see section 1.1), of demonstrable effectiveness [PAR 09].

As intuitive or obvious as it may be for the decision to become a solution to a problem/an answer to a question, the process that leads to it will identify the mechanism that will make it possible to solve the problem posed/answer the question put forward. This solution/answer mechanism involves a number of parameters, including the nature of the problem/question, the prior existence of permissible solutions/answers and the precision of the solution/answer[3]. Because of this, two main approaches can be distinguished, with one or the other being applicable depending on the context. The first approach, "the classical theory of rationality"[4] [VON 44; BOU 06a], advocates the idea of a single and optimal solution to a perfectly defined problem, in conformity with universal laws. Obtained using a formula or a law, such a solution hence becomes independent of the decision-maker's intentions, i.e. two different decision-makers will obtain the same solution to the same problem. As for the second approach, called "the theory of procedural rationality" [SIM 55], it advocates, for a problem that may be incompletely defined, a procedure that allows the achievement of solutions judged to be *at least* permissible, or even satisfactory or favorable. Such solutions are decreed by the decision-maker on the basis of the evolution of the "procedure" type. This procedure is an aid to solving the problem without truly solving it. In this vision, the solution provided will naturally depend on the intentions of the decision-maker, i.e. two different decision-makers could obtain two different solutions to the same problem.

If we were to seek to position one school or the other, we would say that classical theory focuses on the decision, i.e. on the result, on the value maximizing achievement of the objective, while the theory of procedural rationality focuses on the decision process, i.e. the succession of activities leading to the decision, which becomes a value judged satisfactory for the objective. A parallel could moreover be made between this position and the definitions given by the CNRTL (see section 2.2).

On the contrary, the decision called an intuitive decision escapes this framework and leaves room for an evolution of the process which is based solely on the feelings of the decision-maker (see Preface). The decision process in this case becomes "a cognitive process based on the decision-maker's past experiences and emotional data" [BUR 99]. Although it is relevant, we will unwillingly discard this viewpoint, a decision process without analytical reflection, as it opens our reflection to approaches that escape the methodologies and tools that we are seeking to explore and which

---

[3] For the sake of simplicity, we will essentially keep the terms "problem" and "solution" for the remainder of this book.
[4] Many terms are also used to describe this approach: strong rationality, substantive rationality, Harvard school, etc.

are usually used in an industrial environment. So adhering to the "rational" decision, one based on rationality, this chapter will present both the schools of thought cited previously.

2.5. The classical theory of rationality

2.5.1. *The search for a numerical value*

In the classical theory of rationality, a calculation makes it possible to find the best solution to a problem, the rational solution. This calculation may be direct, attached to immediate application of a law. It may also be indirect and obtained following the successive application of a number of laws. It can even happen in some cases and for some types of problem that the best solution may not be absolute and exact but relative and approximate. The decision will then be identified with this solution. The decision process in this case becomes a procedure dictated by the needs of this calculation. The making of the decision will be identified with the translation of the considered objective into the form of a problem, the solution of which conforms to the prerequisites of the calculation. In other words, the challenge of making the decision lies in choosing the solution method, on the one hand, and the mathematical formulation of the problem in conformity with this method, on the other hand; the solution being systematic and the methods defined. Because of this, classical theory hence underlies a need to quantify the elements of this problem numerically and the existence of a mathematical relationship between them. The variable(s) of the system considered (see Preface) can thus be quantified, just like the formulation of the objective (using a function). A relationship between the two could also be established as well as the possible conditions (the constraints) on the values that may be taken by the variables.

Thus, as long as the situation considered can be taken into account in a numerical universe and the idea of a better solution has meaning, it will be the number of variables and the number of individuals or organisms involved that will make it possible to choose the right approach. The decisions that might be made with recourse to this theory are ones whose associated objective is conveyed by an ideal numerical value to be reached, by means of numerical values taken by the variable(s) involved.

Let us concretize, then illustrate the types of decision that could be made within the framework of this classical theory. To do this, and for the sake of comprehension, we begin by considering situations to which this theory is not adapted. As it happens, the problem of improving the quality of working life (QWL) in businesses and the choice of actions to be carried out in this sense cannot be solved within the framework of classical theory. Beyond the qualitative nature of the variables involved (e.g. the mode of management or the ergonomics of the job) and the links between these, finding in this a precise and absolute definition and so a better possible state in effect has little meaning. On the other hand, the shortest path and distance minimization problems, typical of distribution/supply networks, are entirely adapted to the framework of classical theory.

The almost unique role of the decision-maker will thus consist of choosing the method and describing the problem. However, to do this, the decision-maker will need to be able to recognize situations (the object of the decision and the associated objective) that may enter into this framework. For the decision-maker, it is therefore important to guarantee the problem/method suitability and so be able to have answers to the following two questions.

– What are the main approaches of classical theory?

– What are the situations that could be tackled with recourse to this theory?

Seeking to answer these two questions is the subject of the following sections.

2.5.2. *Fundamentals*

Fundamentally, the classical theory of rationality advocates the idea of optimization and is meaningful when mathematics can model the whole solution of the problem posed.

Just like improvement, the idea of optimization is almost intuitive, more or less explained and formalized depending on the context and the decision-maker's intention (see section 1.2.6). For our part, we will keep two main currents of formalization, for their proximity and use in problems of industrial government, i.e. operational research (OR) and game theory.

Concerned by the need to detect German threats during World War II, it was British military logistics that introduced the idea of a better use of resources. They developed methods applying mathematics to allocate limited resources (constraints) as well as possible to land surveillance operations and the organization of convoys. This was the beginning of "operational" research, i.e. research for military "operations" [CHU 57]. Important for engineers and technicians, OR was seen as "the discipline of scientific method that can be used to develop the best decisions. It makes it possible to rationalize, to simulate and optimize the architecture and functioning of production and decision systems" [ROA 21]. More precisely, OR respectively covers programming methods[5] that make it possible to reach exact solutions, i.e. those whose optimal nature is proven, and heuristic methods, which make it possible to reach approximate solutions.

When mechanisms for cooperation between individuals or organisms are put in place, subsidiary questions of interest appear. Introduced in the 1920s [BOR 21] and developed in the 1950s, game theory had the object of proposing solutions to problems considered complex, by seeking to preserve, for a given situation, the interests of each in an interactive, even conflictual context. In this theory, the problem is described in the form of a game involving at least two players, the players being "economic actors" with an interest in line with the situation. Solutions identifying all the values possible for the variables. Benefit or "gain" becomes the objective that each player will seek to optimize (maximize). They will then choose, for the variables involved, the configuration that will make it possible to maximize this gain. Each player will do the same. In doing so, they may gain or lose, depending not only on their choices but also on those of the other players. This theory has some proximity to the strategies developed by chess, go or poker players in which the gain of one player will depend on that of the others. It will moreover be this proximity that gives it its name.

2.5.3. *Operational research*

Table 2.1 summarizes the approaches of OR, depending respectively on:

– the number of objectives (criteria) to optimize;

---

[5] It is interesting to observe that the initial semantics of the word "programming" underpin a semantics of military planning and not a semantics of information science.

– the nature of the solution proposed, whether exact or approximate.

We will discuss respectively single-objective optimization, multi-objective optimization and heuristics. The interested reader will be able to consult [GAR 11] for a complete list of all the associated methods.

	One optimization criterion	Several optimization criteria
Exact solution	Optimization (mono-objective)	Optimization multi-objective
Approximate solution	Heuristic	

Table 2.1. *OR approaches*

When the said objective function, the expression that conveys the objective of the decision, is defined according to a single criterion, the optimization is a single-objective one. Solution methods using programming have been developed in this context to take account of both the nature of the function and the nature of the variables. To each method, there corresponds the development of one or more algorithms allowing it to be implemented in more or less specific conditions. We cite, for example, methods such as:

– linear programming and its famous simplex [DAN 48; DAN 90], when the objective function and constraints are described by linear functions;

– nonlinear programming [CAR 61], when the objective function and the constraints cannot be described entirely by linear functions;

– dynamic programming [BEL 57] and its branch and bound method, when the problem is broken down initially and then optimized a second time.

Let us take the example of the production system of a business creating a portfolio with a number of different products. One question when developing sales and operations planning (S&OP) concerns choice and the quantities to be produced for each product sold. The idea is naturally to maximize the overall profit, or to make the best possible compromise between producing, selling and storage. The objective function will therefore be identified with the sum of the different benefits associated with the marketing of each product. One of the constraints will, for example, identify the limit of the run time for equipment. The application of linear programming will make it possible to determine the optimal production quantities for each product for a maximum profit.

When the objective function is defined according to several criteria, the optimization is multi-objective or has multiple objectives. It therefore means optimizing a criterion vector [GEO 68], on the basis of methods and associated algorithms such as:

– Programming by objectives [CHA 55], in which optimization is successive and tackles the criteria individually, in an order of priority fixed at the outset.

– Compromise programming [ZEL 73], in which optimization involves a comparison of permissible solutions to an ideal solution in which all the criteria would be optimized completely. The least squares method is typical of this approach.

– Lexicographical optimization [EHR 08] that successively restricts, using a mechanism inspired by alphabetical order, the previously known set of permissible solutions until the optimal solution is obtained. The restriction mechanism is based on the search for optimization of a first criterion, then of a second if this is not reached, etc., on as many criteria as are involved.

> A production system meets several criteria, the importance of which varies depending on the context. In the luxury sector, for example, *Product conformity* is vital, *Cost price* less so, and *Work in progress* level[6] still less. Optimization of *Product conformity* first of all becomes the expression of the objective. This non-conformity may be due to the transfer of parts from one assembly station to another. The application of a lexicographical method will make it possible initially to retain, of the permissible solutions to overcome the non-conformity, those that ensure total *Product conformity*. It will then make it possible to retain from these those whose *Cost price* is lowest. If the solution is unique, recourse to optimization of the *Work in progress* level will not be useful, it can be used in the opposite scenario.

Noting the impossibility of defining the optimality of a vector, the concept of a set of optimal solutions is foregrounded [KEE 76], taking up much earlier work, highlighted by Pareto [PAR 09]. In this context, a set of optimal solutions is defined, given that a solution is optimal if there is no solution at least as good, according to all the variables, and better, according to at least one variable. In this sense, an optimal solution is not better than all the others, but no other solution is better than it. To be able to retain a single solution, an exploration of all the solutions can then be considered [FIS 70;

---

6 Work in progress accounts for products undergoing transformation in a production system.

ROY 85], but this time, following an approach that is more in line with the second approach, that of procedural rationality.

> Let us reconsider a production system that satisfies criteria that, this time, have the same importance. This is often the case for the automotive sector, for example. The different permissible solutions will then be described by a vector giving a value for each criterion. Out of the set of optimal solutions in Pareto's sense, a method such as compromise programming makes it possible to decide the optimal solution.

When the optimization problems are considered difficult, with very long calculation times (linked to the enaction of programming methods), recourse to heuristics is preferred. Indeed, the principle of a heuristic or a metaheuristic is to offer a compromise between precision of the solution and the complexity of the calculations. The heuristic is specific to a problem, whereas a metaheuristic is independent of this problem [GLO 86]. We cite, for example, metaheuristics such as:

– simulated annealing [KIR 83], which reasons on the optimum as on a stable thermodynamic state that would be reached by a metallic material after heat treatment;

– genetic algorithms [HOL 75], which reproduce the mechanism of natural selection to evolve an initial solution gradually into a solution close to the optimal one, from a series of cross-breeding, evolution and mutation mechanisms taken from genetics;

– ant colony algorithms [DOR 96], which reproduce the behavior of these insects, the laying down and evaporation of pheromones, to seek the shortest paths.

> For illustration, let us consider a case found in [GAG 01] which addresses the production scheduling of a foundry casting line, i.e. some dozens of manufacturing orders (MO). Among all the orders permitted, the idea is to choose those that minimize the *Loss of casting center capacity*, the *Total delay over the whole order package* and the *Transport capacity losses*, given the constraints linked to production. In view of the complexity of the problem, a metaheuristic such as those of ant colonies makes it possible to come close to the optimal solution in a reasonable time. MOs identify the stages of a path that correspond to their order, and the shortest path would

be the one with the highest level of pheromones. Three orders are proposed: one that minimizes the *Loss of casting center capacity* as a priority, a second that minimizes the *Total delay over the whole order package* as a priority and a third for *Transport capacity losses*. Beyond these rational results, the decision-maker will have to make a choice depending on whether they wish to prioritize a criterion or establish a compromise. This is another situation that takes a look at procedural theory.

2.5.4. *Game theory*

Game theory is a global framework that can be detailed according to a typology of games established in the 1950s [GIR 09]. A game puts forward players and associates them with gains or losses. The most notable game typologies are established according to several aspects.

– The sum: zero-sum games are games in which what one player gains, another player loses. This renders the accumulation of gains and losses zero for the situation considered. In the opposite case, they are called non-zero sum games [VON 44].

– Cooperation: cooperative games are games in which the gain of each player may increase or decrease, depending on whether they cooperate with the other players or not. Cooperating in this case means jointly choosing the configuration to be kept. In the absence of cooperation, the principle of "Nash equilibrium" allows each player to know their optimal configuration, in the sense that they are solutions adopted by the other players, the gain associated with this solution cannot decrease [NAS 50a; NAS 50b].

– Completeness of information: the games involve complete information, where each player knows all the possible solutions they can choose, just as much as all the possible solutions that each other player can choose. Each player is also aware of the gains or losses linked to these solutions as well as the other players' intentions (a search for individual or collective gain, notoriety, etc.). When this knowledge differs between players, it is called asymmetry of information.

– Synchronicity: simultaneous games are games for which the choice of configurations to retain is made simultaneously. In the sequential case, the players choose their solution one after the other, such as in chess, for example.

The example chosen to illustrate this theory is inspired by Cournot duopoly [COU 38; BOU 14]. In a dipole situation, two businesses (such as Airbus and Boeing for example) are the only ones competing on a market that dictates price laws (variations depending on the quantities available). For each business, it is a case of maximizing their gain by putting the adequate *product quantity* on the market. In a context where the offer exceeds demand, this means not inundating the market, and not letting the competition take over the market. The game in place is a zero-sum game, to the extent that gains may jointly increase or decrease. To simplify the problem, we suppose that each business hesitates between two values of the *product quantity* that we describe, for clarity of reading respectively using the terms "little" or "much". Four configurations (and gains) should therefore be envisaged: little-little, little-much, much-little, much-much. Table 2.2 represents these configurations with a simplified illustration of the digital values of the *product quantity* associated with the gains.

Business 2 \ Business 1	Little	Much
Little	Business 1: €€€€ Business 2: €€€€	Business 1: €€€€€ Business 2: €€
Much	Business 1: €€ Business 2: €€€€€	Business 1: €€€ Business 2: €€€

Table 2.2. *Table of gains in a duopoly*

Let us suppose that the businesses do not cooperate in a synchronous context. In this case, where they do not know one another's choices, the maximum guaranteed gain will be that given by applying the principle of the Nash equilibrium. Business 1, for example, would benefit from producing "much", it will be the same for business 2. On the other hand, when the game is cooperative, the best solution is to produce "little" for both businesses, the gain is then €€€€. The business would therefore have a great interest in cooperating.

2.5.5. *Taking account of uncertainty: towards procedural rationality*

The approaches proposed in classical theory were initially developed on the basis of defining both the criterion and the variables involved in a particular (numerical) universe. In fact, there were two conditions underlying the application (see section 2.5.1):

– the parameters of the law linking the criterion to the variables involved are known perfectly;

– the value taken by each of the variables involved is unique and deterministic, for each allowable solution.

In many problems, these conditions are not valid. In fact, it is often projections, estimations or predictions that make it possible to determine the parameters of the law. As it happens, this can be the case for benefits associated with the manufacture of references to catalogued products. This may also be the case for unit times for manufacturing the various components. It may also happen that, for reasons of simplicity, the values taken by variables are approximated by precise digital values, while often they can only be obtained in a different form (a set of more or less certain values). This is the case, for example, in a logistical context, with the basic times involved in minimizing the *Transport duration* for carrying out a journey. This will also be true, in a production context, for the case of defective products that are involved in minimizing the *Product non-conformity rate*, which is often based on mean values (in relation to quantities and products).

Consideration of the random character of some optimization data has pushed thinking towards the use of theories allowing this character to be integrated into the approaches underlying classical theory. Essentially therefore, discussion was about uncertainty and risk. Optimization within this framework makes it possible to provide an optimum that takes not the form of a precise numerical value, but that of an expected value of a random entity, i.e. a set of values compared to a type of average.

The first model dedicated to considering the random character of data was the one offered by probability theory. Each random variable is associated no longer with a single value, but with a set of observed values, to which a distribution of probabilities is associated, as well as an expected value. It is this expected value that is optimized. The first publications involved in this corpus used "objective" probabilities, i.e. ones for which the values taken by the variable, and thus the frequencies, are observable [BER 34]. These publications were then extended [SAV 54] to the case of probabilities called "subjective" probabilities, i.e. ones where since the values (and frequencies) of the random variable are not observable, the distribution of probabilities is constructed according to "probability", i.e. the degree of confidence accorded to the occurrence of the values considered [RAM 31].

> Let us take the example of a business that, when developing its S&OP, wishes to optimize its overall profit by choosing the best product mix in this sense. To simplify, this profit represents the sum of the different profits associated with marketing each product. In turn, the profit of each product is the product of a basic profit with the *Product quantity*. The business knows that these different basic profits are only an estimation proposed by the accounting system. The company knows that the productions are quantified with some lack of precision. One way of overcoming these imperfections is based on probabilities to take considered uncertainties into account, subjective probabilities for modeling the values of the benefits and objective probabilities for that of the *Product quantity*. The values considered will thus be the expected values. From these expected values, it is possible to launch a linear program that will optimize the expected gain. Additional information will be able to strengthen knowledge of the optimal solution, such as gain in the worst or best of cases, respectively the most unfavorable or favorable of the parameters and variables. We might also imagine a framework of gain by calculating the minimum and maximum values for a given probability. Generally, it is the 90% value that is retained since it represents the possibilities almost exhaustively.

The second model, based on possibility theory, which generalizes that of probabilities has discarded the need to identify a probability distribution on the basis of a principle of observing the values and frequencies taken by the random variable. The possibility distribution is based on a construction which relies on an expert interrogation of the "possible" or "necessary" occurrences of the different values of the variables [ZAD 78; DUB 83]. By using this approach, it becomes possible to consider the decision-maker's intention in a vision that exempts itself and which supports the numerical universe as a language for representing and solving problems.

This consideration of the random character of data in classical theory therefore relativizes the result of optimization, since the optimum is uncertain. Thus, in game theory, a single configuration could provide either small or significant gain, and thus poses the notion of risk. This therefore means considering two contradictory trends present in each random game. The first trend concerns risk aversion – "a bird in the hand is worth two in the bush!" – which leads decisions that may be dangerous to be avoided. The second advocates the search for maximum gain – "nothing ventured, nothing gained!" – which leads to risk-taking decisions. Considering these two approaches opens up classical theory towards a consideration of the decision-makers' intentions and makes it bend the principle of systematizing solutions to the problems posed. Modeling of the decision-maker's "preferences" with regard to risk has been discussed [ARR 62; PRA 64].

2.5.6. *A return to the decision process*

After this succinct presentation of the essential parts of the approaches offered by the classical theory of rationality, it is time to return to the notion of the decision process, a central element of this chapter. In order to specify what we see for the three blocks previously identified for this process (Figures 2.2 and 2.3), and in the framework of this theory, we summarize the approaches and main keywords in Figures 2.2 and 2.3.

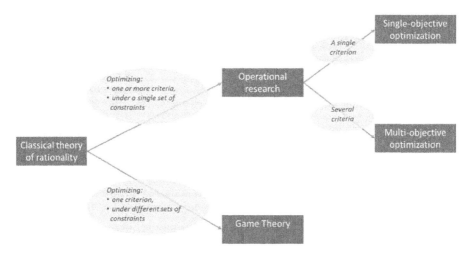

Figure 2.4. *Different approaches of the classical theory of rationality*

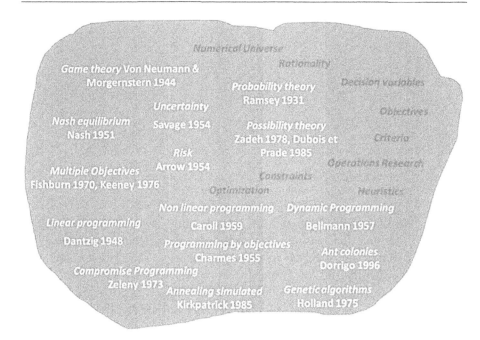

Figure 2.5. *Some keywords from the classical theory of rationality*

In their turn, Figures 2.6 to 2.8 propose a specification of the decision process in the framework of classical theory. In this sense, the process described in Figure 2.6 generally reprises that of Figure 2.3, for which it will specify just a few elements of the three blocks. In this case, the criterion or criteria to be optimized, as well as all the constraints to be satisfied, are listed in the "Before" stage. In the "Decision" stage, the optimization approach for determining the optimal solution is chosen and applied until the values of the problem variables are restored. In the "After" stage the corresponding action plan is defined and implemented. For its part, monitoring will involve execution of the chosen actions with a view to reaching the values defined for the variables considered. It will lead, once the action plan has unfolded, onto the classical stage of expressing the performance. This expression will make it possible, on the one hand, to analyze potential deviations from the values determined at the solution stage and, on the other hand, to take a step back from the action plan and choice of the taken approach.

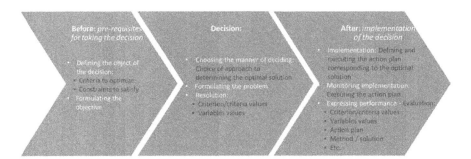

Figure 2.6. *Decision process in the classical theory of rationality*

We note that in this approach, no specification is given concerning the method of defining the action plan or expressing the performance. Moreover, this is not surprising given the subject of the theory. The decision-maker is thus free to identify this performance expression to their own satisfaction with a view to the results reached, or more objectively, to dedicate this expression to the purposes of effectiveness and efficiency [JAC 96]. They might seek, for example, to know at what number the values reached respectively by the (effectiveness) criterion or by the (efficiency) variables are optimal. The decision-maker could also seek to validate the formulation of the problem (objective, constraints), from the choice of approach retained or even from the choice of action plan. Since here again, we can imagine without any difficulty that this action plan may not be unique and that a question of choice may arise, a choice that may not be made within a logic of optimality.

The process described in Figure 2.7 details the previous process, in the case where the approach chosen is the OR approach, providing the specifics belonging to this branch. The approach to determining the optimal solution is in fact specified in the "Decision" stage: exact or approximate optimization, single-objective or multi-objective optimization, certain or uncertain values. As for the "Before" and "After" stages, these still conform to those of the process described in Figure 2.6.

The process described in Figure 2.8 details the process in Figure 2.6 in the case of game theory and remains very similar to the process described before. In the "Before" stage, it is the different set of constraints associated with the criterion considered, that are listed. The approach to determining the optimal solution will be specified in the "Decision" stage, depending on the specifics of the problem and will therefore influence the type of game to adopt: a zero-sum game or a non-zero sum goal, cooperative or not cooperative, with complete or incomplete information, simultaneous or sequential, etc. The activities in the "After" stage remain identical to those of the process in Figure 2.7, concerning the definition and implementation of as many action plans as there are games envisaged, each action plan corresponding to reaching of the values proposed to the variables, specifically to each set of constraints.

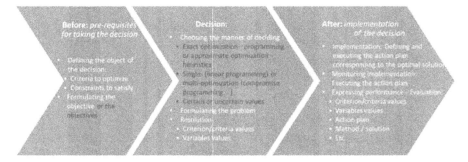

Figure 2.7. *Decision process in operational research*

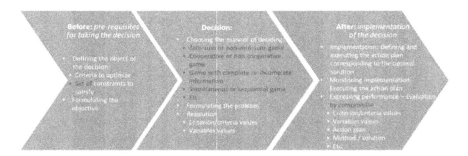

Figure 2.8. *Decision process in game theory*

We can observe that the complexity engendered by such a formulation of the problem can in turn generate complexity in the plans for the "After", just as much in definition and follow-up as in expression of the performance. Faced with this limit of the systematism, only the mechanisms for seeking compromise and negotiation make it possible to reach the different sets of constraints satisfactorily. Since they are not the only one with intentions on the same object, the decision-maker will this time have an additional role, attached precisely to this "dialogue" with the other stakeholders in the same problem.

Thus, we can see both our initial reflections and our naive vision of the decision process confirmed (Figures 2.2 and 2.3). Nevertheless, it is interesting to observe that in this current of thought, it will be the "Decision" block that will be the most decisive. A short-cut is moreover made between the classical theory of rationality and the elements of this block, with identification of the method that allows the solution to be provided to the problem posed as a prerequisite, just like the implementation of the solution given. The decision process becomes almost trivial, given that each approach proposes its procedure and that it is enough to apply the latter to have solutions.

2.5.7. *Summary and analysis*

Deciding on the basis of classical theory underpins a (decision) problem. This problem links the objective of the decision to one or more criteria to be optimized and the system to one or more representative variables of this system. These variables are also significant for reaching the objective. The decision does not only influence how the value linked to this objective is reached but also how this value is fixed. The value to be set for the objective is therefore decided in some ways according to the means available, which are modeled by the variables and the constraints on these variables. The decision becomes a set of numerical variables, which is left to the decision-maker to translate into an action plan that can be executed in the context considered. Often seen as an idea or an extreme, the optimal configuration may in fact be contextualized, forming a kind of reference for what can be carried out.

Many decision cases may be typical of this approach, the examples given in this section and drawn essentially from industrial settings have been one illustration of this. By proclaiming "Whoever has undergone the intense experience of successful advances made in [science], is moved by profound reverence for the rationality made manifest in existence[7]", Einstein in some ways laid down the application conditions for the methods of classical

---

7 https://todayinsci.com/E/Einstein_Albert/EinsteinAlbert-Science-Quotations.htm.

theory. Any situation can in fact conform to it, provided that it has a meaning in which the criterion and variables can be quantified numerically and the link between the criterion and the variables and the constraints on the latter can be formulated.

The classical theory of rationality guarantees the solution since "own logic, purely rational and mathematical, removed from any affective consideration, can lead to a solution whatever the problem" [VAN 08]. It is important to note that in the framework of this theory, anything in the process can be explained since, to quote Einstein, rationality as it is understood in classical theory, "represents the justification for a particular scale of values" [TAT 64]. Consequently, if the objective, or the value(s) linked to the optimization criterion, is not reached, it will not, in the context of applying an approach taken from this theory, be because the "wrong" decision has been made, since the decision is the same regardless of the decision-maker, but because the action plan will have been "badly" defined or the associated action will have been "wrongly" carried out. Indeed, if the choice of method remains the responsibility of the decision-maker, it is not the decision-maker however who really decides but the decision-maker/method pairing, or indeed the method alone. But if the decision problem posed does not fit the mold, wouldn't this in some cases explain the surprising results we might have? Of course, reality will not have been discredited in favor of the application of theory. Variants were introduced to facilitate consideration of real situations, whether from the perspective of the number of criteria to be considered, which may be diverse or antagonistic to one another, of the nature of the constraints and values that may be taken by the variables or the possible generation of a climate of uncertainty or indeed of risk. The whole forms a corpus that bears constant witness to the effectiveness of this framework, as well as its limitations, however few. Reality is not always formed of rationality, determinism and optimization. The method may not exist even though the problem exists.

We will now leave this universe in which the decision is made, rather than to be made and go to see the side where nothing at all is systematic, and moreover, nothing is given. The decision-maker now needs to return to thinking and analyzing… but they will be helped to do this, by the methods of procedural rationality theory.

2.6. Procedural rationality theory

2.6.1. *The search for a satisfactory solution*

Procedural rationality theory involves thinking for oneself about how to solve a problem, the ins and outs of which escape sole recourse to a mathematical approach of the kind seen above. Although the problem posed remains a problem of choosing from among a set of possibilities with a view to reaching an objective (see Preface), it will be impossible this time to draw out, uniquely and irrevocably, the optimal or even the best possibility. "Compromising" thus summarizes the philosophy of this theory.

Procedural rationality theory is present in the absence of a description, indeed a modeling, of a problem in a language that allows for a classical solution, i.e. resolution that leads to a clear solution. The choice of solution will therefore be based on a principle of comparison, by a decision-maker, from among different permissible solutions, so long as these are easily accessed. The chosen solution, in this framework, will be the decision-maker's "preferred" solution, the one they will place above the others, which they will judge to be best adapted to their problem. This solution may also result from their own construction in cases where there is no clear preference. This first underpins the capacity to specify the question "What should we choose from among these possibilities?" in the questions "What possibilities?" and "What should be preferred from among these possibilities?". In second place will come the ability to answer such questions. It is thus that deciding amounts to choosing not the solution that allows unique and optimal reaching of the objective considered (since this is unknown or does not exist), but to draw out the permissible solutions then determine the preferred solution. We might think intuitively that any decision-maker will prefer the solution they find most satisfactory. Indeed, this notion of a satisfactory solution may have many definitions. In fact, it could be defined objectively (the more) as linked to its suitability to the objective, i.e. the more the objective is reached, the better the solution. There again, it may prove in some cases that the expression of this fit may not be trivial. More subjective, the notion of satisfaction might also be linked to what the decision-maker judges to be the most favorable (the best) choice, beyond the fit with the objective. Thus, although the optimum term adopts an absolute and unanimous significance, it will not be the same for these notions of preferable, satisfactory or best. It will be this specificity that will make the problems treated by this theory complex, in which the "free will" of the decision-maker – their knowledge their intention, etc. (see Foreword) – in some ways comes to substitute the previous systematization and determinism more or less completely.

Naturally, if the decision-maker knows what they prefer, i.e. what is satisfactory, then the choice is immediate and does not need theorizing. But in the opposite case, the question they will ask themselves will be the following: "How do we define this notion of the 'most satisfactory' solution and succeed in choosing it?" Answering this double question is the basis of the approaches developed in the framework of procedural rationality theory. These approaches aim to rationalize the process that leads the decision-maker to their choice. The decision will then identify the solution chosen, the one hence judged to be most satisfactory with a view to reaching the considered objective. The decision process in this case becomes a procedure aiming for a structuring or "rationalizing" of the reflection carried out by the decision-maker with a view to making their choice, this structuring starting from declaration of the objective until implementation of the decision that thus corresponds to the most satisfactory way of reaching it. Rationality therefore lies not in the decision but in the process (procedure) which leads to it. Making a decision will be identified with translation of the considered objective into the form of a problem, the solution of which consists of choosing a satisfactory solution from among those allowable or in constructing the latter if this does not appear trivially.

More precisely, procedural rationality theory, on the one hand, underlies the possibility of describing the decision problem in the form of a set of solutions or actions, the respective implementation of which makes it possible to reach the objective. The theory relies, on the other hand, on a solution principle constructed from an idea of comparison "based" on the different solutions put forward. Since any situation can by described by an objective, a set of solutions and a coherent comparison mechanism will therefore find a process in the framework of this theory.

> We will consider again the case of improving quality of life at work (QWL) in businesses, as mentioned in section 2.5.1. The problem posed concerns the choice of an action plan to be carried out in this sense. Unlike classical theory, procedural rationality theory can provide an aid to deciding on a satisfactory action plan so that improving the QWL can be defined by a particular set of action plans comparable to one another. This will involve action plans such as those linked to management methods, to

the definition of working hours, to replacing, storing or handling devices, to the reduction of nuisance noise, etc. It only remains to ensure that this comparison is well founded.

It is not difficult to understand that procedural theory becomes involved when the conditions for applying classical theory are not all met. Faced with a problem of choice, not being able to choose the optimal way requires the decision-maker to have recourse to choosing a satisfactory way. We recall however that, as mentioned previously (see section 2.5.7), in some cases the optimum will not be the choice retained without however calling into question the approach retained to determine it.

Thus, classical theory is well adapted to the problems of determining the shortest paths and minimizing the distance cited previously (see section 2.5.1). There would be little benefit in such cases in substituting procedural rationality theory.

For the decision-maker, making the decision does not therefore amount, in the framework of this theory, to abandoning a previously chosen method. Beyond formulation of the problem in the form required (the object of the decision, associated objective, permissible solutions), the decision-maker will construct this solution aided by a suitable procedure. In this case, the decision-maker will only need to know their problem and be able to compare the permissible solutions put forward. Discovering the procedure for comparison is the subject of the following paragraphs.

2.6.2. *The basics*

"Dear Sir, In the affair of so much importance to you, wherein you ask my advice, I cannot advise you what to decide, but if you please I will tell you how" [FRA 72]. Almost all the basics of procedural rationality theory are implied by this quotation. More precisely, we can infer two main principles from it, i.e.:

– the decision-maker is the only one to make the decision;

– recourse to an approach is possible to help the decision-maker make this decision.

What the decision-maker will judge satisfactory will be specific to them. The first principle reminds us, in this sense, of the weight, in the framework of this theory, of the intention of the decision-maker in making the decision. The second principle, which corrects the potential biases of the first, recommends that this intention may nevertheless by "objectivated", to the extent of using a method to lead the decision-maker to the decision. We add to these two principles a third additional principle, which specifies the adherents of the approach proposed in this theory. The idea is to base this approach on reason. Procedural rationality theory is therefore a rationality considered as being at the same time reasonable, balanced, exact, reflected and sensed [LAR 21; CNR 21]. It proposes, more precisely, decision-making using cause–effect-type cognitive mechanisms. The decision consequently identifies the solution resulting from deployment of this mechanism, symbolized in Figure 2.9.

In classical theory, game theory seeks to take account of the case where several decision-makers are involved in decision processes addressing the same objective. Well-founded optimization approaches however remain the basis of decision-making. Nonetheless, procedural theory goes further and envisages the possibility of a group of decision-makers on the same decision process. The choice of solution satisfactory for the group raises questions of compromise, arbitration, deliberation, etc., which will not be addressed in this work but to which the methods we will state later are adapted.

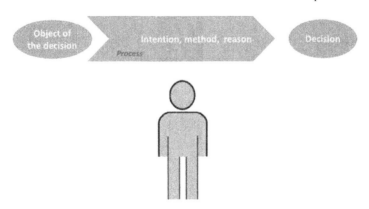

Figure 2.9. *Mechanism of procedural rationality theory*

In complement or in opposition to classical theory, procedural rationality theory will appear more "complex" to understand. Its complexity is explained first by a problem in choosing, but one that this time evades simple determination of a single solution better than all those possible. This thus opens the way to a broader vision of the notion of choice, in the sense that we can choose not only the solution but also "sort" the solutions and "rank" them according to a particular viewpoint, or even simply, "describe" them. Moreover, and as we saw previously, choosing, sorting, ranking or describing may appear beyond a mathematical solution stage, which aims to give an expected value but not the procedure for reaching this value. Essentially introduced by Simon, such a procedure has made it possible to pose the basics of procedural rationality theory and has inspired many approaches and methods.

2.6.3. *The school of Herbert Simon*

It is therefore a question of guiding the decision-maker in their mechanism for choosing, and doing so until one or more solution(s) that they judge to be satisfactory can emerge. The image that came to mind for the father of this theory was one of a chess player, in so far as:

> [...] the expert chess player's *heuristics*[8] for selective search and his encyclopedic knowledge of *significant patterns* are at the core of his *procedural rationality* in selecting a chess move. Third, studies have shown how a player forms and modifies his aspirations for a position, so that he can decide when a particular move is 'good enough' *(satisfices)*, and can end his search [SIM 76].

Procedural rationality theory was proposed formally in the 1950s by Simon. Since then, this theory has provoked an important scientific debate and has generated many publications. Among these, we distinguish in particular those that benefit the domain of decisions in industry directly, in this case the respective work of Mintzberg, Nonaka, as well as Le Moigne.

The economist found that classical theory could not meet needs in the area of decisions and saw three essential limitations to its use, i.e.:

– the imperfect nature of the decision-maker's knowledge of their decision problem, in particular at the level of the declaration of a sometimes misidentified objective;

---

8 "[...] which proceeds by successive approaches, gradually eliminating the alternatives and keeping only a restricted range of solutions tending towards the optimal one" [CNR 21].

– difficulties in anticipating the consequences of the decision made, of the solution chosen, in this case, the value taken by the optimization criterion;

– the potential difficulty of envisaging exhaustively all the actions that are possible to achieve the optimum.

> Let us imagine a context of inventory management. The manager of a store selling supply components feels that their store is not operating properly: some components are lost, others are badly arranged, and there are shortages of some while others have a low turnover. Despite this, this manager does not know, for the moment, how to formulate their objective, for this store, clearly and precisely. Moreover, although they imagine some solutions to remedy this state, such as making a cycle count for the supplied components, a 5S action to level out the current stock, classifying the components according to their turnover, etc., they can only evaluate the consequences of these actions approximately on reaching their objective. Finally, the limited skills of our decision-maker, in the domains of supply management, information systems and industrial organization, limit the solutions envisaged to the technical aspects of managing a store alone.

The reasoning is simple: when the conditions for applying models from classical theory are not met, it is impossible to determine the optimal solution, it becomes opportune to adopt the notion of a satisfactory or effective solution, i.e. a solution that satisfies the needs of the decision-maker [SIM 55]. To do this, it is necessary to reason, i.e. to adopt, in conformity with the principles of Ancient Greece, an intellectual approach based on logic to instruct, describe and understand, and a democratic method based on deliberation in order to choose (see section 1.2.6). Given the impossibility of addressing the problem using formal methods, the approach is therefore based on the cognitive abilities of the decision-maker. This reasoning becomes rational, resulting from a process of structured reflection as characterized by Simon: "Behavior is procedurally rational when it is the outcome of appropriate deliberation. Its procedural rationality depends on the process that generated it" [SIM 76].

The decision process in the framework of procedural rationality is structured so as to conform exactly to the four main stages described above and schematized in Figure 2.10.

– The "Intelligence" stage is intended to perceive a problem, to collect data relating to the system, to make a diagnosis and to formulate the expected objective or objectives. This formulation is also accompanied by all the possible constraints that should be considered.

– The "Design" stage consists of seeking the solution(s) linked to the objectives and constraints expressed.

– The "Choice" stage corresponds to an – initial – selection of the solution that satisfies the decision-maker in view of their objectives and constraints.

– The "Evaluation" stage makes it possible to judge implementation – after the event – and satisfaction linked to the chosen solution.

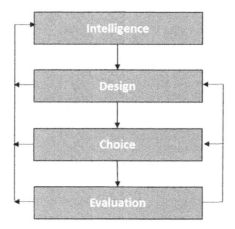

Figure 2.10. *Decision process according to Simon*

We note that the "Evaluation" stage was added later [SIM 60] to describe and compare the solutions designed as objectively as possible, and so to enlighten the decision-maker in the "Choice" stage. Although this process includes four activities, it could have included five, consideration of the "Implementing the chosen solution" stage had in fact been envisaged in this initial version [POM 12]. But this addition will only be used later, as we will see in section 2.6.4. Carried out by the decision-maker sequentially and iteratively, these stages, specified below, allow them to reach a solution that seems most satisfactory to them, and on the basis of an objective that is sometimes merely felt. "Steps back" to earlier stages are permitted, and indeed encouraged. So, for example, the "Design" stage makes it possible to improve the solutions proposed initially after the "Choice" and "Evaluation" stages have made the conclusion that for the time being, no current solution satisfies the decision-maker.

2.6.3.1. *The "Intelligence" stage*

It is in the "Intelligence" stage that the decision-maker becomes aware of their decision problem for the first time, then formulates this a second time. Indeed, they begin by observing, for the system, a deviation between the observed state and the hoped-for state [EAS 73]. This observation is explained by a translation of the observed and hoped-for states into the form respectively of measurements and objectives associated with the system variables involved. The decision-maker then makes a diagnosis on the observed state and on the failure to reach the objectives. This diagnosis will serve as a basis for designing the adequate(s) solution(s) in the next stage. During this stage, the decision-maker will thus seek information both internal and external to the system, allowing them to describe the observed state, to formulate the objectives and diagnose the causes. After this stage, the decision problem is formulated, in the matter of the deviation to be addressed and the objectives to be reached. We note however that this "Intelligence" stage is often neglected in favor of direct design, even though it may prove critical [LEM 73].

> We take the example of inventory management, where the manager knows that they are not meeting the needs of the production lines. There are often recriminations: some components may be missing, the inventory turnover is low and the store is overloaded because of all the components that pass through it (rather than being made available on the production line, for example). By visiting another store in the same group where supplies are managed according to lean manufacturing standards[9], the manager becomes aware of differences in how the two stores operate. They then set objectives in line, respectively, with the *Shortage_rate*, the *Annual_turnover* and the *Load*[10], in this case:

---

9 Lean manufacturing is an approach for improving the performance of production systems, aiming to reduce waste and delays in production.

10 The *Shortage_rate* measures the proportion of missing products expressed as a percentage. The *Annual_turnover* measures the number of times the stock is completely renewed in a year. The *Load* measures the work that must be done by a team and/or an operator over a given period, most often over a week.

– *Shortage_rate* = 0.5% for a value observed for *Shortage_rate* = 3.5%

– *Annual_turnover* = 24 for a value observed for *Annual_turnover* = 10

– *Load* = 50% for a value observed for *Load* = 95%

The diagnosis carried out shows that the deviations are linked mainly to production management favoring a flow strategy and a high inventory level as well as the obsolescence of the information system, dating back more than two decades.

2.6.3.2. *The "Design" stage*

The "Design" stage involves defining solutions to the problem posed. This stage relies on the decision-maker's knowledge and their ability to consult experts to draw up and design solutions making it possible to reach the objectives. This stage may be limited, if applicable, to selecting or adapting existing solutions. This stage is linked directly to the "Intelligence" stage, in the sense that reflection on the solution to a problem is conditioned by the type of problem considered and its causes.

Since each problem is specific, its solution must also be specific, which Simon illustrated, saying that there are no off-the-peg solutions [FRI 04]. In the end, each solution retained (designed or selected) is described according to the selected variables to state the objectives and note deviations. This stage is identified with the states that it makes it possible to obtain in this sense.

To design solutions that correct the deviations observed, the store manager consults with their collaborators. Supply, production and information services are mobilized as well as a logistics consultant and a trainee engineer. They design a good number of "tailored" solutions, covering more or less total reaching of all the objectives considered. These solutions can therefore be combined. These solutions integrate: overhauling the information system in relation to supplies, putting in place an internal logistics using a "milk run" process[11], putting in place a Warehouse

---

11 The milk run combines the supply of components to production stations according to demand and the collection of finished products by logistical methods, always by the same means at a fixed frequency, generally every two to four hours.

Control System[12], replacing MRP planning with a Kanban[13] provider system, putting in place automatic Kardex-type bin storage, etc. For example, the first solution proposed is described as follows:

– *Shortage_rate* = 1.9%

– *Annual_turnover* = 12

– *Load* = 70%

2.6.3.3. *The "Choice" stage*

It is at this stage that the decision is made: the decision-maker chooses the solution that seems satisfactory to them. Based at once on analyses carried out in the previous stages and on the decision-maker's intention, the "Choice" stage, according to Simon, integrates both factual elements and affective elements, which underpins the idea that any decision has both an objective and an ethical sense [SIM 47]. By ethics, the authors mean that the criteria for choosing should include human and moral values that are often implicit: not every solution is a good one to make! The factual aspect relies on describing the future state linked to a given solution. The affective elements rely on the imperative that the solution be effective and align with standards particular to the decision-maker, in other words on how much a factual description of a given solution satisfies them. Although the second aspect relates to the values and feelings of the decision-maker, who combines the notions of emotion, politics, power and personality [MIN 76], notions that are still not accessible to formalization, the first aspect relies on an evaluation. Because of this, the "Choice" stage has no real conclusion, except at the end of the "Evaluation" stage.

For the inventory manager, the choice hinges on the solutions chosen previously. The solution that seems satisfactory again is the one that combines overhauling the information system relating to supplies and the

---

12 A Warehouse Control System makes it possible to manage references, a rolling inventory, location, input and output (potentially in the order of thousands) of stored products, whether supplies, products in the process of creation or finished products.

13 The MRP method (Material Requirement Planning) is a method for managing production, which makes it possible to plan production depending on the demand forecast and the state of the inventories. Kanban is a visual production management method based on labels. The number of labels absent from the production management table represents the inventory level of available products. Production is restarted as soon as the inventory level of products available passes a set threshold that corresponds to exceeding the threshold for the number of labels.

implementation of internal logistic using the milk run process. The store manager will have listened a great deal to the logistics consultant and opted for tried and tested solutions that present little risk. Depending on the results obtained, they will think of finally reconsidering this choice. In this way, they will be able to provide specifications for the solutions retained (sharing supply management with providers, periodicity of the milk run, etc.) or will be able to provide new solutions such as a Manufacturing Execution System (MES). Such a solution will make it possible to know the state of production and inventories in real time and therefore to deduce from this the supplies that need to be made. The chosen solution initially will be described in the following way:

– $Shortage\_rate = 1.5\%$

– $Annual\_turnover = 15$

– $Load = 64\%$.

2.6.3.4. The "Evaluation" stage

Initially confused with the "Choice" stage, the "Evaluation" stage consists of comparing the chosen solution from before its implementation, to the objectives. The result obtained will confirm or invalidate the choice made. Analysis methods similar to multi-objective optimization methods make it possible to provide an aid to the decision-maker in this evaluation, methods that allow the decision-maker to exercise "limited rationality" another name given to this theory and to the work of Simon. By the term "limited", Simon indicates that the search for optimality of classical theory (total rationality) should, when it is impossible to treat the problem with formal methods, leave room to search for satisfactory solutions.

Evaluation of the chosen solution for managing the supply component store is made on the basis of comparison between the hoped for states (the objectives) and those estimated after implementation of the solution that combines overhauling the information system for supplies and implementing an internal logistics using a milk run process. The evaluation is then made by means of calculating the deviations:

– $Shortage\_rate\_deviation = 1.3\% - 0.5\% = 0.8\%$

– $Annual\_turnover\_deviation = 24 - 13 = 11$

– $Load\_deviation = 58\% - 50\% = 8\%$

The solution is not satisfactory, none of the three objectives is reached fully and the deviations remain too great. The "Design" stage is therefore relaunched with two new solutions: one that integrates shared management of supplies with providers (solution A) and another that proposes implementing an emerging MES (solution B). These solutions are evaluated once again by means of calculating the deviations.

Solution A:

– $Shortage\_rate\_deviation = 1.3\% - 0.5\% = 0.8\%$

– $Annual\_turnover\_deviation = 24 - 13 = 11$

– $Load\_deviation = 58\% - 50\% = 8\%$

Solution B:

– $Shortage\_rate\_deviation = 0.5\% - 0.5\% = 0\%$

– $Annual\_turnover\_deviation = 24 - 15 = 9$

– $Load\_deviation = 64\% - 50\% = 14\%$

In view of their intentions to make the running of their store more similar to that of the other store belonging to the company (see section 2.6.3.1), the manager is not satisfied with any of these solutions. They are nevertheless constrained by production management to implement one rapidly to meet the needs of the lines. They decide to make their choice on the basis of a score that they will link to each of the solutions and retain the one with the best score. The scores are calculated by relying intuitively on a combination of the deviations observed for each of the criteria. These deviations will be normalized at first, then averaged a second time. Normalization is achieved by linear interpolation (the value 0 corresponds to the state observed, and the value 1 corresponds to the objective). Thus, for solution A, the normalized deviation corresponding to the *Shortage_rate*, which will be the result of the ratio of the difference between the observed state and the estimated value *after the event* and the deviation between the observed state and the objective:

$$\frac{3.5\% - 1.3\%}{3.5\% - 0.5\%} = 0.73$$

A similar calculation is made for the three criteria and the two solutions. The inventory manager can then average the values obtained:

– Solution A: $\dfrac{0.73 + 0.21 + 0.82}{3} = 0.59$

– Solution B: $\dfrac{1.00 + 0.36 + 0.69}{3} = 0.68$

Where confidence is lacking and in an emergency, solution B will be kept.

2.6.4. *Extensions to Simon's process*

Simon's ideas, formalized in the process described above, have inspired many publications that discuss this process, modify it and complement it, while still retaining the essential idea: a "good" decision is necessarily the result of a "good" process. Among these publications, those of Mintzberg, those of Nonaka and those Le Moigne, whose respective propositions are described in Figures 2.11, 2.12 and 2.13, have excited our interest, as mentioned previously, both for their richness and for their similarity to manufacturing problems.

2.6.4.1. *Mintzberg's model*

Mintzberg's model revisits Simon's process. He extends it both upstream and downstream using the respective stages "Detection", detecting the trigger event (problem or opportunity) and "Implementation", mentioned previously and in which the solution is implanted (Figure 2.11) [MIN 76]. Beyond these extensions, Mintzberg transforms the "Intelligence" stage into a "Diagnosis" stage, for diagnosis and causal analysis. Finally, considering that the "Choice" and "Evaluation" stages cannot always be distinguished, the other proposes a "Selection" stage that covers both and operates by means of three sub-stages.

– The "Judgment" sub-stage is based on the decision-maker's intuition and experience to express the performance, according to reaching of the objectives, of each of the solutions.

– In the "Analysis" sub-stage, the decision-maker introduces factual elements that allow them to characterize each of the solutions.

– The "Bargaining" sub-stage is enacted in the particular case where the decision-maker is surrounded by diverse, involved stakeholders, i.e. when the decision becomes a group decision.

Intended for organizations, this model in particular introduces the mechanism that makes it possible to consider specifically the contribution of a group in a decision process and specify the role of the decision-maker "responsible" for making the decision. This decision-maker should measure the consequences of the choice for each, arbitrate these choices depending on the strategy and argue their decision, thus revealing the role of the manager in a decision process.

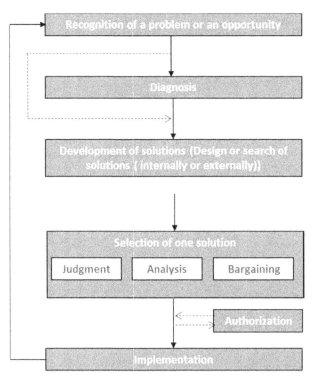

Figure 2.11. *Decision process according to Mintzberg [MIN 76]*

Let us consider a workshop management situation and an overall objective to decrease the *Lead time*. Imagine that management decisions are made for each assembly line by each line manager. They decide on the actions to be carried out in conjunction with a methods technician and an external consultant, as well as the three team leaders.

During the team meeting of one of the workshop's assembly lines, a problem with respecting the *Lead time* was indeed detected ("Recognition" stage). Performance indicators were then updated, and the *Lead time* observed was eight days for a set objective of two days. The diagnostic then made ("Diagnosis" stage) shows that the quantities of work under way were high, the synchronization of posts was poor, the line was poorly balanced and showed a bottleneck in production. Three solutions were then proposed ("Design" stage) by the group:

– implementing a Kanban system to fit the flow of production to demand;

– improving the productivity of work stations by reducing and eliminating unproductive time and in particular setups;

– reducing pressure on the bottleneck by capacity sub-contracting.

Each member of the group then expressed their feeling about the solutions proposed based on the repercussions this would have for them ("Judgment" sub-stage). The consultant thus lent towards the Kanban system, which would ensure they earned consulting fees, the methods technician preferred to automate the least productive machines, thus responding to the directives of the service. As for the team leaders, they wished to improve the capacity of the bottleneck, which would avoid the need for multiple emergency interventions to ensure the production schedule was met. To allow a constructive dialogue on the basis of the different viewpoints, the line manager proposed analyzing each of the solutions in coherence with improving the *Lead time*, given the constraints of the project in the budget and the time limit ("Analysis" sub-stage). On the basis of the description and feelings, an exchange phase was embarked upon by the group members ("Bargaining" sub-stage). The result of this was a new solution, combining the Kanban solution with an improvement in the productivity of the only bottleneck.

After the arguments had been explained by the line manager to the production director, the implementation was authorized ("Implementation" stage). Two months later, the *Lead time* was three days, which satisfied the line manager.

2.6.4.2. *Nonaka's model*

Educated in Japanese and American universities and a student of Simon, Nonaka argues that the explicit knowledge used in the "Intelligence" stage (Figure 2.12) is only "the tip of an iceberg of knowledge [...] which covers, on the one hand, expertise resulting from practical experience" and, on the other hand, "schemas, mental models, beliefs and perceptions" that reflect "our vision of reality (of what is) and our vision of the future (of what should be)" [DEM 05]. Faced with decision problems in the framework of developing new products in companies such as Canon, Nonaka lent two specifics to the decision process. On the one hand, this process is seen as a knowledge creation process. On the other hand, and going beyond the idea of bargaining introduced by the previous model, it is the result of a shared construction (and not that of a single decision-maker) that requires certain conditions be satisfied [NON 91]. Decision is therefore not really at the heart of this model, which instead defines the conditions for success at the "Design" stage. The idea is collectively to animate a working group to make knowledge that was initially implicit and individual become explicit, shared and re-usable. Nonaka consequently structured the decision process (in service of the "Design" stage) into the four interactive stages described below and presented in Figure 2.12 [NON 95].

– "Socialization": the process of converting new tacit knowledge through shared experiences in day-to-day social interaction.

– "Externalization": tacit knowledge is articulated into explicit knowledge... so that it can be shared by others to become the basis of new knowledge.

– "Combination": explicit knowledge is collected from inside or outside the organization and then combined, edited or processed to form more complex and systematic explicit knowledge... The new explicit knowledge is then disseminated among the members of the organization.

– "Internalization": "explicit knowledge created and shared throughout an organization is then converted into tacit knowledge by individuals... This stage can be understood as praxis, where knowledge is applied and used in practical situations and becomes the base for new routines" [NON 08].

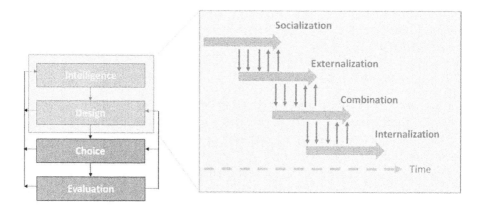

Figure 2.12. *Decision process according to Nonaka [NON 91]*

This vision has been echoed, particularly in concurrent engineering practices. These call into question the sequential character of the process of designing new products [CHA 93] as it was enacted in design offices until the 1990s. The engineering adopted then consisted of a succession of ordered tasks, in conformity with Simon's initial vision, as shown in Figure 2.13.

Figure 2.13. *Sequential design process for new products [DEC 98]*

To summarize, concurrent engineering breaks with this practice by:

– getting stakeholders to collaborate in the design process in a single place, the project board;

– decompartmentalizing the stages of the design process by enabling simultaneous execution of some tasks.

Figure 2.14 describes the design process for new products in concurrent engineering. It favors collaborative tools, as well as exchanges making it possible to construct, gradually and collectively, an integrated solution, the upstream stages having very early knowledge of the feedback and propositions of the downstream stages.

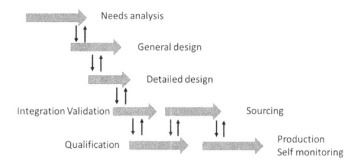

Figure 2.14. *Design process for new products in concurrent engineering [DEC 98]*

This process forms part of Nonaka's vision. In 1986, Nonaka posed the principle of design managed "like a rugby team", by sharing collective knowledge to generate creativity in the game. Such a vision stands in opposition to a "relay race" vision where the players run independently of one another, a vision that had hitherto been favored [TAK 86]. More generally, in the context of Simon's decision process, the idea of concurrent engineering is to improve the performance of the "Design" stage at source, so that the potential solution(s) compared during the "Choice" and "Evaluation" stages may be satisfactory from the first iterations of the process.

> In order to position concurrent engineering, we describe the practice currently carried out by the Facom company, a specialist in hand tools and equipment for mechanics' garages, which developed its new products according to the principles of sequential engineering until the 1990s. At this date, to shorten *Time to market* and reduce *Costs*, it had been decided to adopt the principles of concurrent engineering. In order to do this, the five services in charge of new products (marketing, after-sales, development, industrialization and production) have been regrouped in a single place, the project board, so as to increase the polyvalence of the engineers and technicians and to facilitate communication between them. For each new development project, the corresponding project manager had the task of coordinating and motivating the team that had been formed specifically. The project board allowed everyone to discuss potential solutions, a permanent dialogue was set up for this. The definitive specifications of the new product, which should be drawn up entirely during the analysis phase for sequential engineering needs, had been frozen throughout the project.

> This organization made it possible to increase the number of products by 400%, reduce the time to market by 80%, increase sales by 40%, reduce the design teams by 20% and allocate them, as part of an ethical approach, to other tasks [TIC 76]. These improvements are the direct consequence of the pooling of knowledge from the different services present on the project board and of its collective acquisition.

2.6.4.3. *Le Moigne's model*

Finally, we note the work of Le Moigne which dealt with the specifics of implementing the decision process in complex systems [LEM 94]. In this sense, Le Moigne takes up a typology proposed by Morton according to which decisions are routine and obey standard procedures, or little used and requiring a particular procedure [MOR 71]. Thus decisions may be respectively "programmed", "non-programmed and highly structured" or "non-programmed and weakly structured (or ill-structured)".

A decision is programmed as soon as it is associated with a problem whose solution can obey a pre-existing program, i.e. when it can be tackled with the help of classical rationality theory methods (see section 2.5.3). It then becomes possible to describe completely the variables of the problem and to write the algorithm that is able to find the optimal solution to this problem (structured).

> Wishing to maximize the *Service rate*[14], a workshop manager seeks a form of scheduling that allows this maximization [CHU 89]. Since the number of machines and tasks is limited, this search is typical of programmable decisions for which an algorithm, a pre-existing information program, makes it possible to determine the optimum.

Conversely, a decision is non-programmed as soon as it is impossible to link it to a problem whose solution could rely on a pre-existing program, i.e. an algorithm whose implementation makes it possible to find an optimal solution. Nevertheless, a non-programmed decision is "highly structured" when an optimal solution can be associated with it but cannot be determined, due to complexity or incomparability, for example. An associated optimal solution, similar to the optimal solution, can then be obtained by optimizing "similar" problems.

---

14 The *Service rate* measures the proportion of manufacturing orders (MO) achieved within the planned deadlines.

Let us again consider the scheduling situation. When the optimization (minimization) criteria is, for example, the *Sum of delays*, this problem can no longer be solved by a program as it could be for maximizing the *Service rate*. Even if it is non-programmed, the choice of scheduling allowing this minimization will remain highly structured, with the processing procedure remaining similar to those for known problems. So an initial algorithm will make it possible to select a set of optimal solutions in Pareto's sense, and so reduce the set of solutions to be considered [BAK 74]. A second optimization (post-optimization) stage will propose an optimal solution for a similar problem in the particular case where the tasks to be scheduled have the same duration. Finally, a last stage will make it possible to calculate increased deviation between the optimal solution of the similar problem and the solution of the problem considered (which exists but remains unknown). The workshop manager will then have an associated optimal solution knowing that the maximum deviation between the *Sum of delays* chosen and the optimal *Sum of delays* will remain lower than this increased deviation.

A non-programmed decision is "weakly structured" as soon as knowledge of the similar problem does not allow it to be solved (see section 2.5.3). It then becomes necessary to develop a heuristic specific to the problem. This formalized reasoning generally leads to a satisfactory solution being identified (from the decision-maker's point of view), without reference to the notion of an optimal solution. The heuristic is relaunched until a satisfactory solution has been identified.

In conformity with the decision to put in place an MES to improve the *Shortage_rate*, the *Annual_turnover* and the *Load* (see section 2.6.3), it falls to the information service manager to choose the adapted software package. They hesitate over several propositions, including an internal development. The last investment of this type, the implementation of Enterprise Resource Planning (ERP[15]), dates back more than 20 years, without anyone capitalizing on the associated knowledge. Adopting an agile approach[16], the information service manager, accompanied by a group

---

15 ERP is professional software that manages all the business' functions (commercial, accounting, production, human resources) in an integrated manner.
16 An agile approach is a project management approach based on frequent iteration (every two weeks, for example) of the different stages – design, testing and the deployment of new solutions that mobilize all the actors simultaneously in meetings of limited duration. This type

of experts responsible for specifying the functions expected of the software package, will then:

– write a summary specification that will be made progressively more precise;

– launch a consultation with providers until at least five functionally satisfactory solutions have been obtained;

– develop a draft contract with the providers of these solutions specifying expectations and commitments;

– bring together the business' investment committee to reach a consensus on the choice of provider.

This approach is inspired by heuristics that allow progressive elimination of unsuitable solutions while still refining the project expectations, and doing so until one or more satisfactory solutions emerge before choosing, in a final stage, the most satisfactory solution.

Figure 2.15 summarizes the classification of the decisions proposed.

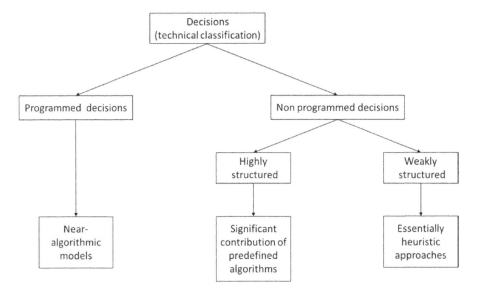

Figure 2.15. *Classification of decisions for a complex system [LEM 73]*

---

of approach allows openness to change, since the needs the project meets may evolve between two iterations, and on the other hand a high interactivity between project actors.

2.6.5. *Procedural rationality and artificial intelligence*

In heuristics, to reproduce human reasoning there is only one step that artificial intelligence (AI[17]) crosses [MIN 59]. Artificial intelligence appropriates this idea of procedural rationality to reach a satisfactory solution for a problem that cannot be solved by the methods of classical theory. It was in this context that Newell and Simon proposed using information technology as well as symbolic reasoning [NEW 58] to solve complex problems. An approach that was called *Human Problem Solving* and which made the hypothesis that human reasoning could be "translated" into computer programming according to a method structured in four stages in the image of procedural rationality [POM 87] (Figure 2.16):

– formulation of the problem considered;

– writing the corresponding program;

– studying the differences between human reasoning and the program;

– using the data to re-write the program and re-iterate the four stages of the method.

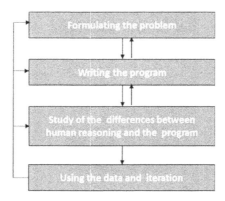

Figure 2.16. *Decision process and artificial intelligence*

The first applications of artificial intelligence in decision-making initially involved decision problems from games, such as chess and tangram, or even control of the first robots in the 1960s. Today, AI involves a broad spectrum

---

17 Artificial intelligence was defined as being the science of making machines do things that would require intelligence if done by humans. It requires high-level mental processes such as perceptual learning, memory and critical thinking [MIN 59].

ranging from medical diagnostics to road traffic regulation, via the monitoring of agricultural crops, activity on the stock market or dialogue with users of applications via the Internet. In industry, it can be used in almost all domains: the quality and traceability of products, predictive maintenance equipment, demand forecasting and production planning, managing product life cycles, etc. This use has been further reinforced today with the advent of Industry 4.0 and its many pillars (augmented reality, additive manufacturing, etc.) [JAY 18].

Even though this wish to compare human reasoning with an information program could be criticized, it nonetheless remains that Simon is considered to be one of the founding fathers of this discipline. Life will have proved him right....

2.6.6. *A return to the decision process*

Just as for classical theory, we will take the time to summarize the different approaches inherent to procedural rationality (Figure 2.17). An essential current of thought thus appears, all or part of which will have been specified for implementation in particular contexts.

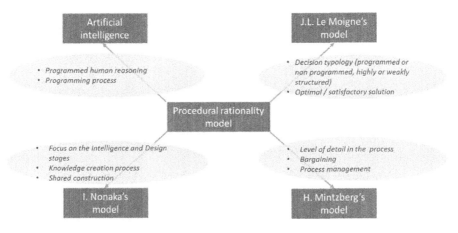

Figure 2.17. *Approaches from procedural rationality theory*

In the same way as for classical theory, Figures 2.18 to 2.21 propose a specification of the decision process within the framework of procedural rationality. The idea of this theory is to focus on the decision process more than on the method, on the solution rather than on the value. Because of this, beyond confirmation of our naive reflections, it is interesting to observe that it is the first two blocks of the process given in Figure 2.3 that are supplied in the framework of this theory. Indeed, the "After" block will not have undergone any particular development in this school. In this case, the "Before" stage corresponds to the "Intelligence of Simon's process. In the "Decision" block that covers the "Design", "Choice" and "Evaluation" stage, it is a reasoned procedure that makes it possible to determine the potential solutions and to evaluate them. This evaluation is carried out with the help of comparison methods with the aim of keeping the solution for which the level of satisfaction suits the decision-maker. If it does not, the process can be repeated at each of the four stages.

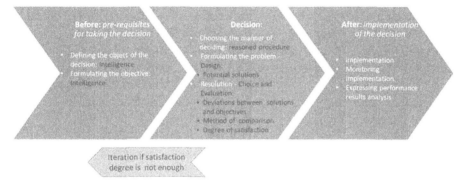

Figure 2.18. *Procedural rationality decision process*

In the case of Mintzberg's model, the process described in Figure 2.19 generally takes up the previous process but adds some specifics, in three blocks. Hence, in the "Before" block, the trigger event for the decision is specified, the stakeholders are identified and a diagnostic, the "reason why" for the decision, is made. The "Decision" block indicates that the "Design" stage mobilizes knowledge that is both internal and external to the decision-maker. This block shows moreover that the "Choice" and "Evaluation" stages are based on the "Judgment", "Analysis" and "Bargaining" sub-stages so as to draw out the satisfactory solution. In the "After" block, the means of monitoring the implementation are defined so as to manage it and evaluate its performance.

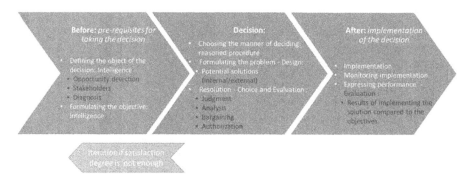

Figure 2.19. *Specifics of the procedural rationality decision process according to Mintzberg*

Concerning Nonaka's model, the process described in Figure 2.20 takes up the process in Figure 2.18 for which he specifies the additions to be made to manage collective knowledge. This translates in the "Before" block into the definition of the working group that will be mobilized for "Socialization" and "Externalization". In the "Decision" block, knowledge management, obtained by "Combination" and "Internalization", allows new solutions to be designed thanks to the pooling of the group's expertise.

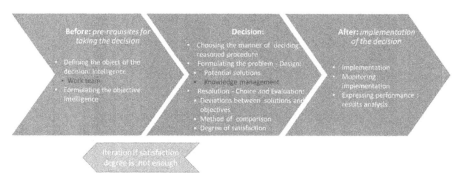

Figure 2.20. *Specifics of the procedural rationality decision process according to Nonaka*

Finally, Le Moigne's model described in Figure 2.21 takes up the process from Figure 2.18 in the case of classifying decisions. The "Decision" block offers the possibility to adapt the problem depending on whether it is programmed, non-programmed and highly structured, or non-programmed and weakly structured. The result is that resolution will give a solution that could respectively be optimal, associated optimal (associated with a similar problem) or satisfactory (see section 2.6.4.3).

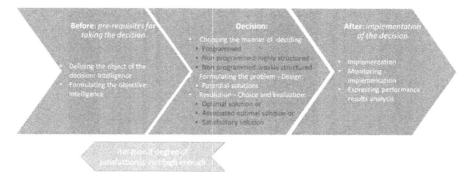

Figure 2.21. *Specifics of the procedural rationality decision process according to Le Moigne*

2.6.7. *Summary and analysis*

Deciding on the basis of procedural theory underlies the choice of one solution from among a set of existing solutions or to designing to answer a problem posed as satisfactorily as possible. The role of the decision-maker in this sense becomes central to the extent that their intentions will impact both the solutions retained and their judgment. In this context, to provide a framework for a decision-maker who seeks rationality, procedural theory proposes not a method but a procedure that the decision-maker is invited to follow to make their decision. This procedure is incarnated in Simon's process, which consists of believing it is necessary to be informed of the problem before designing its solution and that the solution should be chosen after evaluation. A number of cases are typical of this approach, the examples given in this section and drawn essentially from the industrial environment have illustrated it.

Moreover, judging Simon's process to be quite general, publications have sought to specify it, according to the usage contexts. Thus, extensions have been proposed for it to take account of the upstream and downstream stages of making decisions and integrating the collective aspect of some stages of the decision process, or even to create links between these processes and those of classical theory.

Simon's reflections thus had a significant impact and multiple currents subscribe to this search for a satisfactory solution obtained using procedural rationality. Naturally, reflections on decisions continued and still continued within or outside the paths provided by procedural rationality. Nevertheless, the limitations inherent to classical theory have been demonstrated and recorded, thought could and can explore new ways coming from the field of strict mathematics and considering the role of the individual, of the decision-maker in the making of decisions. Here, we cite March, a precursor in the domain of organizational learning, knowledge management or risk aversion and traveling companion to Simon with whom he wrote the reference work for organization theory, *Organisations* [MAR 58]. March illustrates this work by mentioning multiple rationales, in which the decision process is considered systematically. He distinguishes between "*contextual rationality*, in which choice behavior is embedded in a complex of other claims on the attention of actors and other structures of social and cognitive relation, *game rationality* in which organizations and other social institutions consist of individuals who act in relation to each other intelligently to pursue individual objectives by means of individual calculation of self-interest, and *rationality of processes*, in which decisions find their meaning in aspects of the decision-making process and not in the result itself." [MAR 78]. More precisely, "contextual rationality" restricts the field of possibilities by reminding the decision-maker that a decision should be considered within the set of possible decisions for the organization to which they belong. As for the "game rationality", noting that each stakeholder in the decision seeks their own interest, it advocates decisions that should be satisfactory for the organization and locally acceptable for each stakeholder. Both these rationalities form part of a systemic vision of rationality where the decision made is re-situated within the framework of an organization and a "system" of decisions [HAM 05]. Finally, the "rationality of processes" happens through description, whether it is synthetic (the stages of the process) or detailed (the stages are described on as many levels as necessary), of the decision process thus permitting it to be mastered [PIN 86]. It is this rationality that demands the writing of documents establishing quality standards in manufacturing businesses [GAN 16]. There are so many characteristics of the rational mind, but these sometimes need to be summarized.

This theory leaves significant space for humans; it goes without saying that its complexity is of a completely different nature to that seen in the context of classical theory. Indeed, it is when the decision problem is felt but cannot be described precisely, when the data linked to them are known imperfectly, when the potential solutions remain in part to be designed and when another known method does not identify a solution, in other words in all cases where classical theory cannot be applied, when procedural rationality makes it possible to choose a solution satisfactory for the decision-maker on condition that its fundamental principles are respected, in a word, on the conditions that the decision-maker should know what they are talking about and experience what they choose. And it is at this point that the complexity of the approach crystallizes.

2.7. Conclusion

The decision belongs to the decision-maker who tries to make a right decision, hemmed with guarantees. They find them in a quest for rationality that could be given by classical theory or by procedural rationality, depending on the problem posed. Classical theory proposes many paths to propose an optimal solution on condition that the decision-maker has posed their problem properly and has documented it well. When these conditions are not fulfilled, procedural rationality proposes considering the decision as being a process rather than a result. What counts is the "how", the "what" will follow. This process is at once always the same and always different since objectives, context and constraints are never the same from one situation to another. We might therefore speak perhaps of the art of decisions, to choose the right instrument in classical theory and to add to it interpretation for procedural rationality. For the latter, do we not encounter poor decision-makers while others "come out on top" of complex decision problems? But it matters little whether one is an honest artisan or a great artist, Simon's decision-maker is still at the heart of the decision, they make the decision – and are hence literally the decision-maker.

Thus, whether it is made "on a whim" or after "mature reflection", the decision follows a particular number of stages, which are always the same. The idea of this chapter will have been to seek to explain these stages by nevertheless distinguishing decisions, in relation to which the decision-maker has little control from those that they form and evaluate themselves. The first category of decisions is based on systematic methods that guarantee a single solution to a problem posed. The second category concerns problems that cannot be solved other than through analysis, reflection, classification, sorting, etc. For this last category, the result is no longer an optimal numerical value, but one "alternative" among others, judged, according to the decision-maker, to be more satisfactory than the others.

Chapter 3 presents the principle of the methods, called decision-aiding, that make it possible to enact the decision process for decisions in the second category. It consists of defining the notions and vocabulary belonging to this domain focusing on the particular case of decisions in the presence of multiple criteria. It presents in particular the notions of comparison, order and preference, allowing the decision-maker to judge what alternative(s) are most satisfactory.

3

The Decision: The Multi-criteria Universe

3.1. Introduction

Deciding means choosing. This choice concerns "the" solution – alternative – to a problem previously posed in terms that call for it to be solved. Formulated in this way, the verb "to choose" may appear to acquire an unusual synonym: to solve. More precisely, if the solution to the problem posed exists and is unique, it will be determined immediately and the choice will then make sense. This choice becomes "systematic", not dependent on the decision-maker, when a method exists making it possible to decide the solution univocally. Nevertheless, there are situations for which application of such methods cannot be envisaged, because of the presence of a set of potential solutions or indeed problems, the reality of which differs from that of a model that lends itself to this. In such cases, determining the solution will require a pathway and will become the result of a mechanism based on the decision-maker's knowledge and intent, and involves "reflecting, discussing, organizing and, finally, choosing" (see Preface). By relying on Simon's process or one of its variants (see section 2.6.4), the decision-maker will then be aided to rationalize the mechanism that allows them to carry out this analysis. This is the essence of what we have retained from Chapter 2.

Although the process of procedural rationality makes it possible to link an approach for structuring the different stages that allow the choice in question to be reached, it does not, however, provide the tool or the formula that might indicate this choice, unlike classical theory. In fact, although we know how to choose – without hesitation – the solution that is better than the

---

For a color version of all the figures in this chapter, see www.iste.co.uk/berrah/decision.zip.

others in all respects, we do not know how to choose – without hesitation – the solution that is "partially better" than the others. The many "partly adequate solutions" in this sense become subject solely to the decision-maker's evaluation. The decision-maker, so as to be able to choose one solution, will be led to express, as far as they know how to and are able to, their preference and their satisfaction with regard to each of them. In this sense, several theoretical contributions have been developed, providing methods that are usable, scientifically well-founded and whose results can be explained, to come to the aid of the decision-maker in the different stages required by the choice incumbent on them.

In continuity with the philosophy of procedural rationality, this chapter focuses on a particular framework in which tools and methods to help the decision-maker have been developed: the framework of multi-criteria decision-aiding (MCDA). In this direction, we will begin, during our intuitive analysis of the notion of choice, by eliminating the case of a decision called a single criterion decision, for which the choice is often similar to a systematic decision. Indeed, the postulate considered here is twofold:

– if the choice cannot be a systematic one, this means there is some complexity in the problem;

– the complexity of the problem results in the need to take into account several dimensions linked to it, since these dimensions are more or less redundant and convergent.

Continuing our investigations relating to the notions of decision and decision-maker, the goal sought at this level of analysis (section 3.2) is to understand the basics of the MCDA framework and the place it gives to the decision-maker, before, lastly, considering its methods. A brief summary on constructing MCDA (section 3.3) will make it possible to start to understand its substance. On the basis of the vocabulary used in this school (section 3.4), suitable notations and formalizations will then be introduced as well as a return to the notion of order (section 3.5), on the basis of the mechanism of choosing. The notion of preference and the nuances formed in this current will be presented before a proposition of modeling by suitable order. The particular case of Pareto dominance will also be addressed (section 3.6). This discussion will be in view of the benefit that, in certain situations, the notions of dominant and dominating solutions could have for managing to isolate not the solution, but the preferred solutions. The statement of MCDA principles will be supported throughout the chapter with industrial illustrations centered on a situation involving implementation of a software program for managing a workshop. But for the moment, as a sort of preamble, let us return to this notion of choice, which we made synonymous with comparison.

3.2. Comparing to be able to choose

3.2.1. *Intuitive vision*

In its most general sense, the notion of choice involves a set of at least two elements in which a kind of selection is carried out in view of the problem posed. According to CNRTL (*Centre national de ressources textuelles et lexicales* – National Center of Textual and Lexical Resources), choosing consists of: "Taking someone or something in preference to another because of its qualities, its merits, or the esteem in which it is held". Because of this, it involves foregrounding one preference among the elements considered in view of "its qualities, its merits, and its esteem", which we see, not absolutely, but relative to its correspondence to the problem at the root of the choice. This preference can only be obtained by comparing these elements. When this comparison is easy, the development of the preference and therefore the choice are easy too. When it is not easy, the development of the preference and therefore the choice, are not easy either. Intuitively, we can imagine that this comparison is easy when it is based on a single "key" for comparison – provided that this allows comparison – which is a single criterion. On the contrary, the comparison becomes less easy when it is based on several keys of comparison, or several criteria.

It will therefore be the number of "criteria" involved in the comparison and therefore in the choice that will account for its complexity. For precision, CNRTL defines a criterion, from the Latin *criterium* meaning judgment, as being a "character or property of an object (person or thing) according to which a judgement is made". By criterion, we therefore understand the variable that influences this judgment. In this sense, the preference between two elements depends on the value taken by the criteria that allow these elements to be described.

3.2.2. *In a single criterion universe*

Let us take an example in which the information system manager is tasked with choosing an MES (Manufacturing Execution System) software (see section 2.6.4.3). There are abundant choices available, as some tens of providers offer this type of program software. After visits to trade shows, consultations with peers and after reading specialist reviews, the information system manager initially collects general information on a dozen offers and selects those they judge to be most suited to the needs of their system. In line with their company's investment policy, which encourages, as French public procurement demands, choosing the cheapest offer based on the *Acquisition_price* to compare the different propositions. They therefore choose the solution for which this criterion has the lowest value: the cheapest MES.

Thus, as long as information is available on the elements to be compared, and as long as the comparison criterion is also available and allows for comparison, we come very close to the notion of "systematic" choice, mentioned previously and based on the presence of methods that allow this to be achieved. Nevertheless, there is the possibility of all things being equal among the elements. If the choice, regardless of one or other of these elements, does not suit, recourse to additional criteria would be necessary.

3.2.3. *In a multi-criteria universe*

In informal conversation with the production team, the information system manager realizes that the choice of MES cannot be based solely on the *Acquisition_price*, but that it involves other criteria, such as:

– the *Response_to_specifications*, which gives information on how the solution conforms to the need;

– the *Associated_service*, which concerns the aspects relating to the integration, training, maintenance and updating of the solution;

– the *Technical_feasibility_of_implementation*, which characterizes the solution's suitability to the installations available on the production site;

– the *Collaboration_level*, relating to the provider's ability to develop functions specific to the customer's needs throughout the lifecycle of the solution.

The information system manager therefore returns to their catalogue. They nevertheless choose to consider only the six providers selected previously and contact them to obtain detailed information on their solutions, in conformity with the set of new criteria chosen. They therefore find that it is impossible to prefer one solution "wholly" to the others.

And it is in such cases that the decision-maker, such as our information system manager, will need help in being able to identify the element they would prefer in order to make the choice. Indeed, situations may appear not only where all things are equal but also situations of incomparability, because of satisfactions that diverge according to different criteria. These may leave the decision-maker baffled. The decision-maker will therefore need to put in order not only their ideas but also the solutions offered to them[1].

Order between elements is thus the key to choosing and thus to deciding. The MCDA current, mentioned previously, centers these propositions on creating this order. MCDA proceeds by order, in turn, to do this: characterizing, preferring, then ordering. The vocabulary is specific to it and rigorously defined, although offering some slight divergences depending on which school of the current is used. But is it not up to the decision-maker to color their decisions with their intentions and culture, were they to choose the definitions inherent to the decision? We will remain in this register for now, of going back to the origins of MCDA, certainly we will better understand how it was formed.

3.3. The construction of MCDA

MCDA was formed gradually and still continues to be developed today [FIG 05a]. To illustrate the concept in general, we cite Roy who, inspired by procedural rationality, was one of the first to suggest a framework for it: "Decision aiding is the activity of the person who, through the use of explicit but not necessarily completely formalized models, helps obtain elements of responses to the questions posed by a stakeholder of a decision process. These elements work towards clarifying the decision and usually towards recommending, or simply favoring, a behavior that will increase the consistency between the evolution of the process and this stakeholder's objectives and value system" [ROY 85].

---

1 In this chapter we previously supposed that these solutions are known by the decision-maker.

The main sources of MCDA were, respectively, social choice theory, for its solution methods (see section 1.9), and the theory of procedural rationality for its process (see section 2.6.2).

In particular, among the approaches developed in the framework of social choice, those relating to the following factors are retained:

– election procedures;

– the "supreme good" of a community and the "social well-being" of individuals.

Indeed, in the 1780s, to respond to election problems (in which it was a question, for an assembly of voters, of choosing one representative from a list of candidates), de Caritat, Marquis of Condorcet, and de Borda, both passionate about mathematics and politics, each proposed a voting procedure. One advocated a relative vision that was based on comparisons per pair of candidates [CON 85], whereas the other adopted a more absolute vision, centered on the ranks obtained by each candidate [DEB 81; MAS 00]. Related to our context, the candidates become the elements and the voters become the criteria. The decision-maker, the people in the original context, is in this case found applying a systematic procedure over which they have no control.

Moreover, during the same period, Bentham then Mill took up the notion of utility in Bernoulli's sense [BER 34], for the purposes of maximizing "community well-being", understood as the benefit of a community, i.e. a set of individuals. For the authors, utility served to designate what contributes to the well-being of a community. We therefore speak of calculating the utility of an action, which is greater the more it tends "to augment the total happiness of the community; and therefore, in the first place, to exclude, as far as may be, every thing that tends to subtract from that happiness" [BEN 89]. The "community well-being" was then the result of aggregating the different utilities envisaged. Related to our context, the actions become the elements and the collective well-being reflects the aggregation of criteria. Here too, the procedure is systematic, similar to that of classical theory. In this movement, at the start of the 20th century, the economist Pareto defined as optimal any situation in which it was impossible to increase the "social well-being" of an individual without reducing that of one (or more) individual(s) [PAR 09]. A situation is then called optimal in Pareto's sense if it meets the previous condition, and not optimal (in Pareto's sense) in the contrary case. Related to our context, situations become the elements and the individuals become the criteria. There too, the specified model systematizes the procedure for choosing, with a vision of choice extended, however, not to the selection of a single element but to a classification of elements (as optimal/non-optimal).

In the 1950s, pursuing the quest for representativeness for voting procedures, the work carried out by Arrow had made it possible to draw out three of these properties that guaranteed their capacity to provide a ranking representative of voter choice [ARR 62]:

– unanimity: when all voters chose the same candidate, this candidate is ranked first;

– the absence of a dictator: the ranking should not depend on the choice of a single voter;

– independence: the ranking of one candidate compared to another should not depend on the votes for the other candidates.

Arrow had, moreover, shown that unanimity, the absence of a dictator and independence could never be respected simultaneously. This was the introduction of the famous Arrow theorem [ARR 98] which meant that no voting procedure could represent voters' choices perfectly. So much for social choice theory.

It was from these two centuries of thought and suggestions on voting procedures and the work of Simon on the possibility of recourse to an approach for aiding the decision-maker to make their decision (see section 2.6.2) that the first suggestions would appear for using formal tools for deciding on solutions suitable for the problems posed [FIS 67; DEB 83]. Appearing as a form of evidence when the conditions for applying the models of classical theory are not met, the decision aid was born and could be identified by its English acronym, MCDA.

So, in the 1960s, inspired by Condorcet's voting procedure Roy developed a family of methods, called outranking. The idea was to avoid a mathematization of optimization methods that was judged to be excessive and to reproduce human reasoning as far as it is possible to do so [ROY 68a]. It is this approach that is identified with what is called the

European school of decision aids, which today includes more than 350 researchers[2] [ROY 97].

Moreover, in the 1970s, taking up the ideas of Bentham on the mechanism for aggregating individual benefit, methods known as aggregation methods were developed. The idea was then to synthesize "community well-being" into a single value, called overall utility, from the decision-maker's preferences, according to the criteria chosen for the solutions considered. This approach was identified with what was called the American school of decision aids, and relies on MAUT (Multi-Attribute Utility Theory) and MAVT (Multi-Attribute Value Theory) [KEE 76, DYE 05].

MCDA, with its two schools, can be seen as a crucible benefitting the results of classical theory and procedural rationality theory. It has resulted in the emergence of a new vision for decisions. Still very young, it is the natural place for a scientific comparison between its two schools, linked to the human and the social sciences, on the one hand, and to the engineering sciences, on the other hand. That said, rich and diverse in its content, the MCDA current relies on a specific vocabulary that is introduced in section 3.4.

3.4. Vocabulary

In MCDA, a "decision problem" is posed as soon as the "solutions" and "criteria" are defined.

3.4.1. *Solutions*

Solutions are generally described as actions or alternatives in the MCDA literature and, more rarely, as objects or options. The European school links the notion of solution to that of action. In this, we find the idea that the solution, which ultimately identifies the decision made, is seen as the culmination, the concretization of the decision and so of the action that results from it (see Preface). More precisely, the authors of this school define the notion of action as "a representation of a possible contribution to the overall decision that is likely, given the state of progress of the decision process, to be envisaged autonomously and to serve as an application point

---

2 For this reason, this community acquired a group within the EURO (Association of European Operational Research Societies) in 1975, see: http://www.cs.put.poznan.pl/ewgmcda/.

for the decision aid" [ROY 85]. As for the notion of alternative, much used in the English language literature, this obeys the same definition as that of action, however with a condition of exclusivity, in the sense that the solution chosen is necessarily unique (two solutions could not be retained even in a situation where all other things are equal) [BEL 02].

The vision adopted in this book advocates the achievement of a single solution; it will therefore be the term "alternative" that will be retained, considering the shared understanding of the notion of choice in its exclusive aspects and its corollary renunciation (see Preface). In our view, having several solutions addresses a slightly different problem that could be conveyed not by the question "What solution should be chosen?" but by the question "What solutions address the problem?". This distinction, also made in MCDA, has made it possible to identify different problems that could be understood using methods from this current of thought, broadening the understanding of choice. So, if it happens that two or more solutions are preferred in the same way by the decision-maker, this could have several implications, such as redundancy between them, a lack of information about them, indifference towards them, indeed, the inability to decide (see Preface). For its part, although the industrial terrain presents as many situations of exclusive "choice" as situations of "selection", it nonetheless remains that this selection often serves an ultimate choice.

> The information system manager is faced with six solutions that would each more or less suit their needs in the matter of MES. These six solutions are identified by their commercial designations, respectively: *XETICS LEAN*, *AQUIWEB*, *ORDINAL COOX®*, *Worderware QUASAR* and *CREATIVE CUBE*. Formulated in MCDA vocabulary, these solutions become the following alternatives:
>
> – acquisition of MES: *XETICS LEAN*;
>
> – acquisition of MES: *AQUIWEB*;
>
> – acquisition of MES: *ORDINAL COOX®*;
>
> – acquisition of MES: *Worderware*;
>
> – acquisition of MES: *QUASAR*;
>
> – acquisition of MES: *CREATIVE CUBE*.

By characterizing all the alternatives of the decision problem posed, the decision-maker starts their decision process. They lay the first brick. The choice of these alternatives will, however, remain marked by their intention (see Preface and section 2.6.1). Thus, two different decision-makers may not choose the same alternatives for the same decision problem.

> During initial consultations on the project, the production manager indicated that they favored an extension of ERP SAP[3] functions currently used by the business. The IT (Information Technology) manager suggested continuing to develop internal solutions, which corresponded perfectly, according to them, to the operators' needs. Responsible for all actions linked to the business' information system, the information system manager nevertheless stuck with their intention to acquire an MES, in coherence with other projects developed in the business.

Thus, we assume the existence of m ($m \in N$) alternatives a_j ($j = 1,..., m$), to the problem considered. We will write all of these alternatives A, $A = \{a_1,...,a_m\}$. When a pair of alternatives is considered, it will be written (a_j, a_k) with $j,k = 1,...,m$. When the number of alternatives is low, it will be simpler to have recourse to a simplified notation that would identify the first alternative by the letter a, the second by the letter b, etc.

> The alternatives chosen previously for acquiring an MES can be written as follows:
>
> – a_1: XETICS_LEAN;
>
> – a_2: AQUIWEB;
>
> – a_3: ORDINAL_COOX;
>
> – a_4: Worderware;
>
> – a_5: QUASAR;
>
> – a_6: CREATIVE_CUBE.

---

3 SAP (Systems, Applications and Products for data processing) is the dominant ERP on the market today for program software for integrated management (in the industrial environment).

3.4.2. *The criteria*

MCDA proposes a vision of the notion of criterion that may be surprising at first if we consider the previous definitions (see section 3.2.1). Beyond the fact that this notion has a meaning with regard to element, solution and now alternative, its definition relies in this case on that of a similar notion, the attributes, which relates to a characterization of an element, whatever that may be. In MCDA, the attribute is sometimes called a "view point" and is identified with a variable, making it possible to describe the alternative considered, whether qualitatively or quantitatively [GRA 05a]. The idea will then be that the decision-maker considers as many attributes as needed to describe all the alternatives envisaged for their problem. This description is generally presented in the form of a table called a "performance table", in the sense that a performance is associated with each alternative and in view of each attribute.

> According to the information system manager, the five attributes enabling a complete description of the alternatives with regard to acquiring an MES are:
>
> – v_1: *Acquisition_price*;
>
> – v_2: *Response_to_specifications*;
>
> – v_3: *Associated_service*;
>
> – v_4: *Technical_feasibility_of_implementation*;
>
> – v_5: *Collaboration_level*.

Thus, we assume the existence of n $(n \in N)$ attributes v_i $(i=1,...,n)$ associated with an alternative a_j $(j=1,...,m)$ of the problem considered. The mechanism that links the performance p_{ij} to a_j according to v_i, is made according to the function v_i following:

$$v_i : A \rightarrow P^i$$

$$a_j \rightarrow v_i(a_j) = p_{ij}$$

where P^i is the set of performances p_{ij} taken by the attribute v_i for a_j. Table 3.1 presents the form of the performance table for a set of m alternatives and n attributes.

Attributes Alternative	v_1	...	v_i	...	v_n
a_1	p_{11}	...	p_{i1}	...	p_{n1}
...	...	...	...	...	...
a_j	p_{1j}	...	p_{ij}	...	p_{nj}
...	...	...	...	...	...
a_m	p_{1m}	...	p_{im}	...	p_{nm}

Table 3.1. *Performance table for m alternatives and n attributes*

The information system manager now seeks to characterize the different alternatives chosen according to each of the five attributes defined. The performance table given in Table 3.2 presents all of these characterizations. In this case and, for example:

– the *Acquisition_price* (established by the editor for use on 20 posts with a service guarantee lasting 10 years) relating to MES *XETICS LEAN* is 9,280 €: $v_1(a_1) = p_{11} = 9{,}280$ €;

– for the *Associated _service*, the characterization is less immediate. It in fact involves trying to evaluate the quality of the service provided by each offer. To do this, the information system manager meets with the workshop manager as well as the IT manager to evaluate satisfaction with the level of service proposed, taking account, as specified before, of the integration, maintenance and updating of the software and training employees in its use. The workshop manager proposes relating this satisfaction to the time needed to cover the different stages, whereas the maintenance manager instead leans towards considering the difficulty and complexity generated. For their part, the information system manager proposes referring to the different functions in the specifications and evaluating the alternatives depending on how they meet these. They therefore propose defining, more generically, five satisfaction levels, i.e.: *Not_very_satisfactory, Fairly_satisfactory, Satisfactory, Very_satisfactory, Extremely_satisfactory*.

In this sense and, for example, analyzing MES *XETICS LEAN* has shown that only basic functions were proposed, which therefore conferred on it the *Not_very_satisfactory* level.

Attribute Alternative	v_1	v_2	v_3	v_4	v_5
a_1	9,280 €	4	*Not_very_satisfactory*	*Very_satisfactory*	*Not_very_satisfactory*
a_2	34,000 €	5	*Very_satisfactory*	*Not_very_satisfactory*	*Satisfactory*
a_3	13,660 €	7	*Satisfactory*	*Satisfactory*	*Very_satisfactory*
a_4	10,890 €	6	*Satisfactory*	*Very_satisfactory*	*Satisfactory*
a_5	15,000 €	4	*Very_satisfactory*	*Fairly_satisfactory*	*Satisfactory*
a_6	16,500 €	6	*Not_very_satisfactory*	*Fairly_satisfactory*	*Not_very_satisfactory*

Table 3.2. *Performance table linked to choice of MES*

Through this introduction of attributes, the decision-maker is helped in reasoning through their decision procedure, to the extent that they are guaranteed an identical analytical framework for all the alternatives. By bearing their problem in mind, the decision-maker first begins by reflecting on all the magnitudes that may characterize the alternatives considered. At the same time, they take care to provide a homogeneous characterization of them. Secondly, the decision-maker allocates values to these magnitudes, still within the logic of decreasing the impact of any subjectivity, to "do the same thing for each of the alternatives". Everything is rational or almost so, since just as for the preselection of alternatives, preselection of the attributes still falls to the decision-maker. In this sense, two different decision-makers may not consider the same attributes for the same alternatives, nor characterize them in the same way. Everyone has their own decision problem in MCDA!

Focused on monitoring the performance of production lines, the production manager suggests to the information system manager to consider the *Performance_monitoring* function as an additional attribute. Moreover, also retaining the *Associated_service* attribute, the production manager proposes quantifying this on the basis of execution times. Although they listen, the information system manager does not take on these propositions, which they consider inherent to an MES's specifications.

On the basis of its identification with the "character or property of an object" (a person or thing) according to which "a judgment is passed" (see section 3.2.1), the notion of criterion has this in common with the notion of an attribute, that it represents a characteristic of the (alternative) element considered. In MCDA, although the attribute sticks to the idea of qualifying this characteristic, whether qualitatively or quantitatively, the criterion conveys the additional semantic of a capacity to establish a preference among the alternatives to which it relates [ROY 05]. The criterion therefore becomes, in this current, synonymous with the values that it takes, and underpins a sort of correspondence between these values and the decision-maker's preference. This is easily understood for criteria such as the *Product-cost*, for example, or its *Delivery_time*, for which the preference increases with the decrease of one or the increase of the other, respectively.

This vision led to the following formal definition: "A criterion is a function g, defined on [the set of actions], taking its values in a totally ordered set, and representing the decision-maker's preferences according to some point of view" [VIN 92]. By mentioning the notion of point of view, the author enables us to emphasize the decision-maker's role in this quantification procedure, which attaches the returned values by the function of a magnitude, a characteristic, a variable. As defined, the criterion thus identifies the point of view – the intention – of the decision-maker both in the choice of magnitude and in its quantification. We also feel the slight confusion that may occur between terms depending on schools of thought. Moreover, we will see in section 3.5.2 why the notion of preference takes on a plural meaning in this sense.

The attribute–criteria transformation will remain a prerequisite for applying the methods proposed.

We assume the existence of n ($n \in N$) attributes v_i which characterize an alternative a_j ($j = 1,...,m$). We assume moreover that to each attribute there can be associated a criterion (which is generally the case in MCDA). The g_i function formulated above identifies the mechanism that transforms a given attribute v_i into a criterion, i.e. links to the value of the attribute p_{ij} the value of the criterion c_{ij} depending on the following g_i function:

$$g_i: \quad P^i \to G^i$$

$$p_{ij} \to g_i(p_{ij}) = c_{ij}$$

where G^i is the set of values taken by c_{ij}[4]. An alternative c_{ij} is thus described according to the set $C = \{c_{1j}...c_{nj}\}$ $(i=1,...,n)$ and $(j=1,...,m)$ of n criteria.

This transformation is generally given in a table of "results", in the sense that a result is associated with each alternative and in view of each criterion. Table 3.3 gives the form of this table, for a set of m alternatives following a set of n criteria.

Criterion Alternative	g_1	...	g_i	...	g_n
a_1	c_{11}	...	c_{i1}	...	c_{n1}
...	...	...	...	...	...
a_j	C_{1j}	...	c_{ij}	...	c_{ni}
...	...	...	...	...	...
a_m	c_{1m}	...	c_{im}	...	c_{nm}

Table 3.3. *Table of results for m alternatives and n criteria*

There are many ways of defining the g function. Without being exhaustive, we can cite approaches based on analytical linear interpolation methods [JAC 82; VAN 86] or normalizations. We can also mention methods based on threshold definitions (of indifference or of preference)

---

4 Some authors directly link the value of the criteria to the alternatives without recourse to the value of the attributes. The value taken by the criterion in this case for an alternative a_j is given by the g_i function, which is then defined as follows: $g_i: A \to G^i$; $a_j \to g_i(a_j) = c_{ij}$.

[FIG 05a] or indeed those based on exhaustive comparison of alternatives [SAA 77]. This being the case, the decision-maker can completely ignore the nature of this function g applied to their problem.

Naturally, the transformation caused by the g function will not be necessary in all cases.

> The information system manager first takes the *Acquisition_price* as the basis for choosing their MES. This attribute becomes a criterion to the extent that it makes it possible to compare the different software offers to one another: the lower the value is, the better the MES corresponding to this point of view. The information system manager will thus prefer the *XETICS* offer with $p_{11} = 9,280$ €, to the *ORDINAL COOX*® offer with $p_{13} = 13,660$ €

On the other hand, in the case where the attributes are defined on domains that do not make it possible to link an order, the g transformation is needed.

This is particularly the case when the best value taken by a quantitative attribute is neither the smallest nor the greatest. Thus, for an attribute such as the *Lot_size* of a production line, a low value can cause too many series changes and degrade productivity, while a high value can increase the production time and degrade reactivity. The best value will result from a compromise between reactivity and productivity. Recourse to the g function is also justified when it comes to making the criteria commensurable and thus easier to interpret.

> When they consider the *Associated_service*, the information system manager has no difficulty in establishing an order between the satisfaction levels *Not_very_satisfactory*, *Fairly_satisfactory*, *Satisfactory*, *Very_satisfactory*, *Extremely_satisfactory*. However, a numerical transcription of this order can make it even easier to interpret. It then appears fairly simple to opt for a characterization in the form of the rank of each alternative depending on this criterion. The g function will respectively associate the ranks:

– 1 to MES *AQUIWEB* and to MES *QUASAR* which have the highest satisfaction level with the value *Very_satisfactory*;

– 3 to MES *ORDINAL COOX* and to MES *Worderware*, which have a lower satisfaction level with the value *Satisfactory*;

– 5 to MES *XETICS LEAN* and to MES *CREATIVE CUBE* which have the lowest satisfaction level, with *Not_very_satisfactory*.

Table 3.4 presents the table of results for choosing an MES. Criteria g_1 and g_2 are numerical values and criteria g_3, g_4 and g_5 are ranks, deduced from the respective satisfaction levels of the various alternatives.

Criterion Alternative	g_1	g_2	g_3	g_4	g_5
a_1	9,280 €	4	5	1	5
a_2	34,000 €	5	1	6	2
a_3	13,660 €	7	3	3	1
a_4	10,890 €	6	3	1	2
a_5	15,000 €	4	2	4	2
a_6	16,500 €	6	5	4	5

Table 3.4. *Table of results for choosing an MES*

Identifying the criteria through the results table will thus form the second stage in the decision process.

Moreover, if we are allowed to restrict the decision semantics to a semantics of choice, the MCDA broadens the outline to additional semantics, addressed in section 3.4.3.

3.4.3. *Decision types*

Deciding means choosing… and even a little more in MCDA. Deciding will not only mean choosing, but also ranking, sorting and scoring [TRI 00; ROY 05]. More precisely, the MCDA takes two different forms:

the decision, on the one hand, and decision-aiding[5], on the other hand. The decision will be identified with the choice, whereas decision-aiding will be seen as "a formal and abstract approach" [TSO 06] that, by processing the available information, "makes it possible better to analyze, understand, explain and justify a problem and/or a solution" [TSO 06], thereby providing not prescriptions – the alternative to be chosen – but recommendations, a set of solutions, an order, a classification, a characterization. For its part, the choice consists of keeping the *Satisfactory* alternative, on the basis of a comparison/elimination principle operated on all the alternatives (e.g. recruiting a new collaborator, acquiring software (the example considered), investing in equipment, etc.). Figure 3.1 illustrates this type of situation for which the alternative a_4 is retained from the set $A = \{a_1, a_2, a_3, a_4, a_5, a_6, a_7\}$.

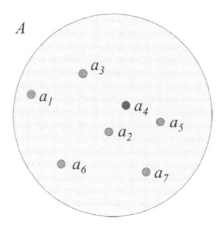

Figure 3.1. *Illustration of choice in MCDA*

As its name indicates, the ranking is the instantiation of an order relationship applied to the alternatives, whether this means ranking from best to worst, with the potential for all things being equal, according to the decision-maker's preference, in view of the problem considered. The ranking is typical of production planning and the ordering of problems, for example, in which MOs (manufacturing orders) are ranked by priority, customer waiting times, manufacturing times, etc. Figure 3.2 illustrates this type of

---

5 The discussion carried out previously (see section 3.4.1) concerning the terms "action" and "alternative", moreover, comes from introducing this notion of a decision aid.

situation where the rank k of the alternative a_j is written $r(a_j) = k$ ($k \in \{1,2,3,4,5,6,7\}$) for all alternatives $A = \{a_1, a_2, a_3, a_4, a_5, a_6, a_7\}$.

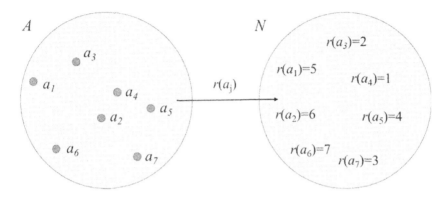

Figure 3.2. *Illustration of ranking in MCDA*

Beyond its original goal, the ranking makes it possible to "prepare" for choosing or sorting. As for sorting, it consists of ranking each alternative in a class among a set of predefined, disjunct classes, ranked according to the desired semantic. The problem of sorting is a classical one in inventory management (ABC method)[6] or in risk analysis (FMECA)[7], or indeed in skills management, where the idea is to group the alternatives in conformity with a profile. Figure 3.3 illustrates this type of situation using three classes, numbered according to their rank, for all the alternatives $A = \{a_1, a_2, a_3, a_4, a_5, a_6, a_7\}$.

Attached to the problem posed, scoring, unlike the previous categories, consists of quantifying the set of alternatives and criteria relating to the problem considered. Going further than merely posing the problem, scoring offers additional information on the overall score linked to each of the alternatives. Seen in the MCDA current as a whole problem, we can, however, observe a direct link between this scoring option, advocated by the American school (see section 3.3), and the choice, ranking or sorting that can

---

6 The ABC method makes it possible to rank the references stored in three categories valued as strong, medium and weak [CHA 06].
7 Failure mode, effects and criticality analysis (FMECA) makes it possible to identify the risks of failure in a system and the means to overcome them [CHA 06].

be deduced from it. In this sense, scoring is widely adopted by companies, where the quantification reassuring the decision-makers with its supposedly objective character tends to be imposed whatever the nature of the decision problem. Figure 3.4 illustrates this type of situation to which the score of alternative a_j is written $U(a_j), j \in \{1,2,3,4,5,6,7\}$ for the set of alternatives $A = \{a_1, a_2, a_3, a_4, a_5, a_6, a_7\}$.

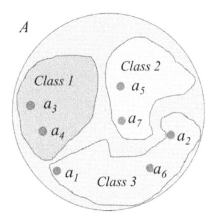

Figure 3.3. *Illustration of sorting in MCDA*

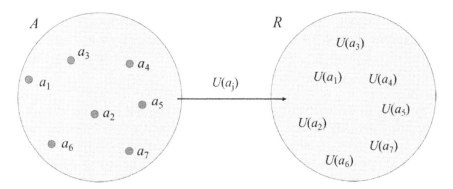

Figure 3.4. *Illustration of scoring in MCDA*

Finally, MCDA offers a framework that makes it possible to process information describing the alternatives depending on the criteria, so as to better understand and facilitate choice. It may mean choosing one alternative from a number, ranking the alternatives from the most satisfactory to the least satisfactory, categorizing them according to how they correspond to the profile of a class or quite simply allocating them a score. The reader will have understood that these problems fit with one another. For example, the ranking can prepare a choice or a sorting, which makes it valuable and we venture this adage: whoever can rank can choose, whoever can rank can sort. Moreover, scoring makes it possible to process any of the three other problems. This being the case and, as almost always, everything rests on the decision-maker and their intention: they identify the alternatives that potentially meet their objective, they define attributes and criteria to compare them, and, if it is not suitable for addressing their problem, they choose an MCDA method to achieve it. But, ultimately, if there are several decision situations, obtaining an order of alternatives will remain a necessary condition for understanding them.

3.5. Ordering for comparison

3.5.1. *Intuitive vision*

Alongside the equivalence relationship, the order relationship, generally visualized in the so-called Hasse diagram [VOG 95], can structure the elements of a set. CNRTL defines an order as being a "a relationship of succession, a sorting obeying a rule or convention". An order relationship over a set thus makes it possible, in conformity "with a rule or a convention", to compare the elements of this whole. It makes it possible, because of this, to obtain an order, depending on the associated preferences, between these elements. Applying an order relationship to a set amounts to characterizing each element by rank. The elements are ordered, for example, "from smallest to largest", "from lightest to darkest", "from oldest to most recent", etc. The order finds its instantiation in the ranking category described above (see section 3.4.3).

The orders developed over a set of elements will not necessarily be identical. In all cases, the result obtained will include a greater or lesser proportion of strict preferences, i.e. allowing a total order between the elements considered. It will be this proportion that will make it possible to specify the order obtained. Identical situations (two elements presenting the same size, the same color, etc.) or situations where it is impossible to compare (expensive but very comfortable vehicles vs. less expensive but less comfortable vehicles, etc.) between elements may in fact appear. This will only allow, given the state of knowledge, development of weak or partial orders. Such orders will prompt the decision-maker to introduce more elements and more rules, or potentially to prioritize the latter.

In general, the postulate chosen in MCDA is the possibility that the order relationship identifies the decision-maker's preference, with a sort of identicality between the type of order obtained and the initial preference. Order in this framework may involve alternatives just as much as criteria. Ordering the alternatives is intended to obtain the preferred alternative. Ordering the criteria is intended to define their relative importance. It will be the third stage in the decision process, once the alternatives are selected and the criteria introduced.

The order relationship models the decision-maker's preference. How is this preference expressed in MCDA?

3.5.2. *The notion of preferences*

CNRTL defines the notion of preference as being a "judgement of esteem or feeling of predilection by which a person or thing is given pre-eminence over another". Preference is a relative notion that thus sheds light on the intentions of the decision-maker with regard to their objective and the means of reaching it (see Preface). The MCDA current takes as its own this vision of preference and its link to the decision-maker, and models it by a suitable order type. A decision-maker with a preference for one alternative rather than another will choose it rather than the opposing alternative. However, this preference can take several forms. Three possibilities will be offered to the decision-maker:

– preferring one of the two alternatives;

– preferring neither of the alternatives;

– not being able to prefer one alternative to the other.

These three situations identify what are called preferences (see section 3.2.2) in the plural, respectively: "strict preference" in the first situation, "indifference" in the second and "incomparability" in the third. The strict preference results from the existence of a total order between the alternatives, whereas indifference and incomparability result in an absence of order between them.

> Depending on the *Associated_service* criterion selected for acquiring an MES, the information system manager encounters two situations to compare the alternatives (Table 3.2) with, in this case:
>
> – a strict preference for the alternative of acquiring a_2: *AQUIWEB* to that of acquiring a_1: *XETICS LEAN*;
>
> – indifference between the respective alternatives of acquiring a_4: *Worderware* and a_3: *ORDINAL COOX®*.

Naturally, the problem becomes more complex when the decision-maker is faced with more than two alternatives and the number of criteria increases. They may have, for each of the criteria, a strict preference over the set of alternatives while not knowing how to state this same preference when they consider all the criteria. Situations where it is impossible to compare alternatives, or indeed situations where there is indifference are generally observed, prompting the decision-maker to go further in processing the knowledge they have. More precisely:

– one alternative is strictly preferred to another if it is strictly preferred according to at least one criterion and indifferent according to the others;

– one alternative is indifferent to another if it is indifferent depending on all the criteria;

– one alternative is incomparable to another if it is strictly preferable to it according to at least one criterion and this other alternative is strictly preferable to it according to at least one criterion.

> When the information system manager considers all five criteria chosen for acquiring an MES, to which they allocate the same importance (for the moment), they encounter strange situations to which they do not know how to respond.
>
> This is the case with the three situations presented above.
>
> – The information system manager has a strict preference for a_4: *Worderware* compared to a_2: *AQUIWEB* according to three criteria, a strict preference for a_2 compared to a_4 according to one criterion and

indifference according to one criterion. What can they conclude about their preference between two alternatives?

– The information system manager has a strict preference for a_4: *Worderware* compared to a_3: *ORDINAL COOX®* according to two criteria, a strict preference for a_3, compared to a_4 according to two criteria and indifference according to one criterion. What can they conclude about their preference between the two alternatives?

– The information system manager has a strict preference for a_2: *AQUIWEB* compared to a_1: *XETICS LEAN* according to three criteria and a strict preference for a_1, compared to a_2 according to two criteria. What can they conclude about their preference between the two alternatives?

The answer to these three questions is the same: the strict preferences and indifference expressed according to each of the criteria do not make it possible to conclude on an order between these alternatives, taken two by two. As it stands, the alternatives are hence incomparable. By carrying out this work over all the alternatives, no conclusion on the preference of one compared to another is possible.

In this third stage of the decision process, the decision-maker finds themselves expressing their preferences. This way, in MCDA, of quantifying all information leaves little room for subjectivity. It is highly probable that two different decision-makers may express the same preferences faced with situations described by values with a sense of "better than" or "less than". It will have been the stage of passing from attributes to criteria that will have allowed this objectivation of a notion that is linked strongly to the decision-maker. The rationale of the procedure will have been striking at this level.

The order relationship models the decision-maker's preference.

Finally, we see the proposition of the MCDA current to do this.

3.5.3. *Preferences and order relationships*

Modeling preferences by an order relationship amounts to giving a rank to each of the alternatives in view of its preferences:

– $r_i(a_j)$ is the rank of an alternative a_j according to the criterion g_i;

– $r(a_j)$ is the rank of this same alternative a_j according to the set of criteria (see section 3.4.3).

> The order obtained by considering the *Acquisition_price* criterion that the information system manager wishes to minimize (Table 3.4) is described in the Hasse diagram[8] in Figure 3.5 where $r_1(a_j)$ is the rank of the alternative of alternative a_j according to the *Acquisition_price* criterion. It appears that:
>
> – alternative a_1: *XETICS LEAN* is strictly preferred to all the other alternatives, $r_1(a_1) = 1$;
>
> – alternative a_4: *Worderware* is strictly preferred to all the other alternatives, except alternative a_1: *XETICS LEAN* $r_1(a_4) = 2$;
>
> – alternative a_3: *ORDINAL COOX®* is strictly preferred to alternatives a_5: *QUASAR*, a_6: *CREATIVE CUBE* and a_2: *AQUIWEB*, $r_1(a_3) = 3$;
>
> – alternative a_5: *QUASAR* is strictly preferred to alternatives a_2: *AQUIWEB* and a_6: *CREATIVE CUBE*, $r_1(a_5) = 4$;
>
> – alternative a_6: *CREATIVE CUBE* is strictly preferred to alternative a_2: *AQUIWEB*, $r_1(a_6) = 5$;
>
> – alternative a_2: *AQUIWEB* is not strictly preferred to any alternative, $r_1(a_2) = 6$.
>
> The information system manager only has strict preferences between the alternatives. The order is total in this case (see section 3.5.1).

---

8 In this diagram, the nodes represent the alternatives and the arcs the strict preference relationships. To simplify the representation and in view of the relationship's transitivity property, only the arcs between elements of immediately successive rank are indicated. When the alternatives are of the same rank, they form a new node that encompasses them.

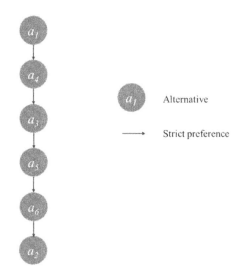

Figure 3.5. *Total order between alternatives according to the Acquisition_price*

The order obtained by considering the *Associated_service* criterion that the information system manager wishes to classify by decreasing satisfaction (Table 3.4) is described in the Hasse diagram in Figure 3.6 where $r_3(a_j)$ is the rank of the alternative of alternative a_j according to the *Associated_service* criterion. It appears that:

– alternative a_2: *AQUIWEB* is strictly preferred to all the other alternatives, $r_3(a_2) = 1$;

– alternative a_5: *QUASAR* is strictly preferred to all the other alternatives except alternative a_2: *AQUIWEB*, $r_3(a_5) = 2$;

– alternatives a_4: *Worderware* and a_3: *ORDINAL COOX*® are indifferent, $r_3(a_4) = r_3(a_5) = 3$;

– alternative a_6: *CREATIVE CUBE* is strictly preferred to alternative a_1: *XETICS LEAN*, $r_3(a_6) = 5$ and $r_3(a_1) = 6$.

The information system manager has only strict preferences and indifferences between the alternatives. The order is partial in this case (see section 3.5.1).

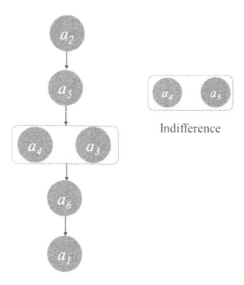

Figure 3.6. *Partial order between alternatives according to the Associated_service*

To observe the two diagrams in Figures 3.5 and 3.6, we can easily see a difference between the type of order associated with the strict preference situation and that (or those) associated with the other preferences. What is valid for one criterion is also valid for many. As such, as mentioned above, complexity increases when there are multiple criteria.

For simplicity, we will leave the information system manager and their proposed acquisition of an MES for now; we will pursue our illustration of the relationship through a purely academic example, since the goal at this stage in the book is only to characterize the problem and not to solve it.

To illustrate the different types of order (total or partial), let us imagine three decision situations for which the decision-maker successively considers a set of three alternatives $\{a,b,c\}$, then adds a fourth alternative $\{a,b,c,d\}$ and finally adds two new alternatives $\{a,b,c,d,e,f\}$. In the first situation, the decision-maker considers the subset of alternatives

$\{a,b,c\}$. Analysis of the table of results, including the corresponding Hasse diagram is presented in Figure 3.7, makes it possible to draw the following conclusions:

– the decision-maker prefers alternative a to all the other alternatives, according to all the criteria, $r(a) = 1$;

– the decision-maker prefers alternative b to alternative c, according to all the criteria, $r(b) < r(c)$.

The decision-maker therefore only has strict preferences between alternatives. The order is total in this case.

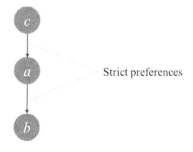

Figure 3.7. *Total order for the subset* $\{a,b,c\}$

In the second situation, the decision-maker considers the subset of alternatives $\{a,b,c,d\}$. Analysis of the results table, including the corresponding Hasse diagram, is presented in Figure 3.8 and makes it possible to draw two new conclusions:

– the decision-maker prefers alternative d to alternative c, according to all the criteria, $r(d) < r(c)$;

– the decision-maker is indifferent to alternatives b and d, according to all the criteria, $r(b) = r(d)$.

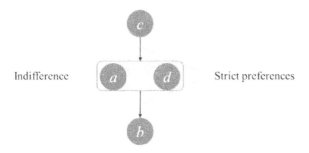

Figure 3.8. Partial order for subset $\{a,b,c,d\}$

For the third situation, the decision-maker considers all the alternatives $\{a,b,c,d,e,f\}$. Analysis of the results table, including the corresponding Hasse diagram, is presented in Figure 3.9 and makes it possible to draw the following new conclusions:

– the decision-maker prefers b to e and d to e, and this, according to all the criteria, $r(b)<r(e)$ and $r(d)<r(e)$;

– the decision-maker prefers e to f following one criterion and expresses indifference between the two according to another criterion, $r(e)<r(f)$;

– the decision-maker prefers c to e according to one criterion and e to c according to the other criterion; they can therefore conclude neither a preference nor an indifference between these alternatives, $r(b)<r(e)$ or $r(c)>r(e)$ or indeed $r(c)=r(e)$;

– the decision-maker prefers c to f according to one criterion and f to c according to the other criterion; they can therefore conclude neither a preference nor an indifference between these alternatives, $r(c)<r(f)$ or $r(c)>r(f)$ or indeed $r(c)=r(f)$.

The decision-maker therefore has strict preferences, indifferences and incomparabilities between the alternatives. Just as for the previous case, the order is only partial.

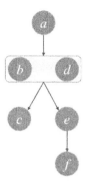

Figure 3.9. Partial order for the set $\{a,b,c,d,e,f\}$

The different illustrations presented above refer to the three types of order considered in MCDA, which are:

– total order;

– total pre-order;

– weak pre-order.

It is called total order, as well as complete, simple or linear order, when the preferences are exclusively strict between the elements of the set considered. For any pair of alternatives, two cases can be presented:

– a is strictly preferred to b: $r(a) < r(b)$;

– b is strictly preferred to a: $r(a) > r(b)$.

Moreover, it is called complete pre-order when the preferences between the elements of the set considered are strict preferences, or indifferences. For a pair of alternatives (a, b), three cases can then be presented:

– a is strictly preferred to b: $r(a) < r(b)$;

– b is strictly preferred to a: $r(a) > r(b)$;

– a is indifferent to b (and reciprocally): $r(a) = r(b)$.

Finally, it is called weak pre-order in the simultaneous presence of strict preferences, indifferences and incomparabilities between the elements of the set considered. For a pair of elements (a, b), four cases can then be presented:

– a is strictly preferred to b: $r(a) < r(b)$;

– b is strictly preferred to a: $r(a) > r(b)$;

– a is indifferent to b (and reciprocally): $r(a) = r(b)$;

– a is incomparable to b (and reciprocally): nor $r(a) > r(b)$, nor $r(b) > r(a)$, nor $r(a) = r(b)$.

The respective total and weak pre-orders are responses to what we have called partial order. Among the three types of order, total order allows the decision-maker a quasi-systematic choice, because there is only one alternative on the first level. However, a complete pre-order or a weak pre-order potentially leaves the decision-maker faced with a subset of alternatives from which they must choose on the basis of elements other than those used to express their preferences.

It can easily be understood that in these conditions, the decision-maker can look for methods able to provide a total order. Recourse to the preferences of the decision-maker alone does not make it possible to guarantee the establishment of such an order; it will be necessary to apply to the performance table additional processing based on integrating new knowledge (the "rules or conventions" seen previously in section 3.4.1).

3.6. The particular case of Pareto dominance

The concept of Pareto dominance (see section 2.4) finds its application in the presence of a large number of alternatives, the preferred of which does not stand out. Initially, the Pareto principle addressed the consequences of improving the "social well-being" of an individual on the same "social well-being" of each of the other individuals (see section 3.3). The idea was that this "social well-being" reached its maximum as soon as its improvement triggered the breakdown of another's.

In a decision context, the consequence to consider is the possibility "of improving" an alternative according to one criterion without degrading this alternative on each other criterion. In other words, no identified alternative exists that can be at least as good as the alternative considered according to all the criteria and better according to one criterion. In this case, where the alternative reaches its "maximum", it is said that it is not dominated by any other alternative. However, if an identified alternative is not better than the alternative considered according to any criterion, while still being less good according to at least one criterion, this alternative will be called a dominated alternative. It is easy to conclude that the decision-maker will not have to consider the dominated alternatives in their choice. They can then reduce all the alternatives to consider non-dominated alternatives, also called satisfactory alternatives. These alternatives form the so-called Pareto front.

> When the information system manager was faced with the question of acquiring an MES, there were so many possibilities (more than 12 propositions) that they had to reduce the number of propositions to a smaller set. They therefore sought initially to dismiss all the propositions that did not present any benefit for them. In this instance, some offers with the same profile as others while still being less advantageous according to one or more criteria (they are more expensive, for example) will not have been chosen. To identify these less advantageous offers, the decision-maker compared, exhaustively, each proposition to all the others, according to all criteria.

Still for the sake of simplification, the notion of a Pareto front is illustrated through an academic example, described below.

> We consider a set of 26 alternatives, written $\{a_1, a_2, ..., a_{26}\}$ and each represented by a colored circle. Each alternative is positioned by its values according to two criteria, one on the x-axis and one on the y-axis. For both criteria, the decision-maker's satisfaction is decreasing. For each pair of alternatives, the question posed is: Is alternative a_j ($a_j \in \{a_1,...,a_{26}\}$) dominated in Pareto's sense by another alternative a_k ($a_k \in \{a_1,...,a_{26}\}$ and $k \neq j$)?

The alternatives for which the answer is "the alternative is not dominated by any other alternative" therefore form the Pareto front. The graphical representation proposed in Figure 3.10 is the counterpart of a results table, which would be more complex to develop.

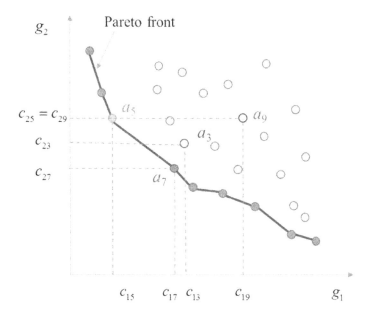

Figure 3.10. *Illustration of the Pareto front*

Since processing all of these alternatives is quite time-consuming, we detail the latter for the subset $\{a_3, a_5, a_7, a_9\}$ which makes it possible to illustrate the different cases that may be encountered. In this case:

– a_3 is dominated by a_7: $c_{17} < c_{13}$ and $c_{27} < c_{23}$;
– a_9 is dominated by a_5: $c_{15} < c_{19}$ and $c_{25} = c_{29}$;
– a_5 is not dominated by a_7: $c_{15} < c_{17}$ and $c_{25} > c_{27}$;
– a_7 is not dominated by a_5: $c_{17} > c_{15}$ and $c_{27} < c_{25}$.

The comparisons are carried out exhaustively. All of the undominated alternatives, as is the case for alternatives a_5 and a_7, therefore belong to the Pareto front.

Thus, initially faced with a set of 26 alternatives, the decision-maker already knows that their choice will involve one of the nine alternatives of the Pareto front.

3.7. Summary

Three stages of the decision process have been carried out up until now:

– defining (selecting, restricting) the alternatives;

– defining the criteria;

– developing the order between the alternatives.

The MCDA current takes the stages defined previously, renames them and enriches them. These are summarized in Figure 3.11, in conformity with Belton and Stewart's process, a reference in the MCDA current [BEL 02].

The "Problem" structuring stage, in this process, involves defining the alternatives and identifying the criteria to which the designation of the stakeholders participating in the process is added. It also involves describing the behavior of the decision environment.

The "Buiding the model" stage identifies the stage of defining the criteria and the method for eliciting their value to order the alternatives.

Aside from constructing the order of the alternatives, the stage "Using the model to inform and challenge thinking" has the goal, as its name indicates, of providing information (sensitivity and robustness analyses) that can call the established order into question. Moreover, it offers the possibility of proposing new alternatives developed from analyzing the "weaknesses" of the alternatives initially chosen.

At these three stages, MCDA adds a "Developing an action plan" stage that poses the question of "after" the decision but leaves the decision-maker at liberty to choose how to achieve this "after".

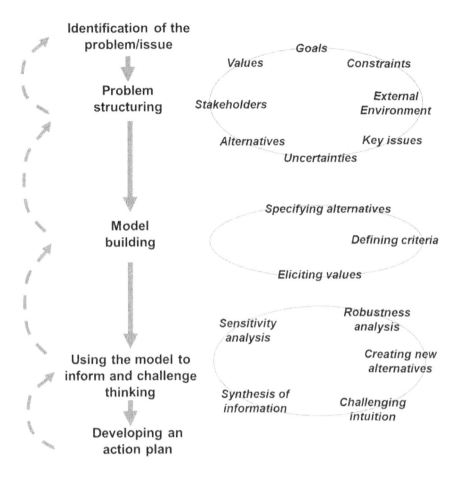

Figure 3.11. *The MCDA process according to [BEL 02]*

We will now revisit this process with regard to the model proposed in Chapter 2. In conformity with Simon's vision (see section 2.6.3), this process is merely a more detailed and more specific rewriting of it, as were the revisited processes we saw before (see section 2.6.6). Indeed, the previous "Problem" structuring stage belongs in the "Before the decision process" block. The "Building model" and "Using the model to inform and challenge thinking" stages belong generally in the "Decision" block. The cyclical character of the process (see section 2.6.6) (repetition of the "Before" block) is triggered by the stage "Creating new actions" when none of the alternatives hitherto considered satisfy the decision-maker. The

"Developing an action plan" stage, on the edge of MCDA discussion, corresponds to the "After" block.

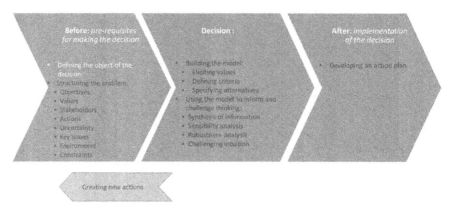

Figure 3.12. *The MCDA decision process (Belton's vision)*

3.8. Conclusion

To decide is to choose, to choose is to compare, to compare is to order. This is the thread running through the ideas presented in this chapter. In this sense, the intrinsically linked notions of preference and order have been defined, commented upon and applied. The whole idea of the decision aid is there: subscribing to a process, running through its stages to define the alternatives and then the criteria, and finally developing an order between the alternatives, most often incomplete however, which leaves the decision-maker faced with indifferent or incomparable alternatives. But what progress if we recall our decision-maker who, at the start of this chapter, was contemplating their alternatives without the power to conclude which they should retain or even imagine!

By retaining this idea of building a total order, our decision-maker has not had their final word. They can undoubtedly express more preferences and give them nuance to refine their criteria; they can ponder them so as to reduce the number of incomparability and indifference situations and so facilitate choice. But, careful not to do it "any old how", our decision-maker knows they will have to advance methodically to construct and use this new information correctly. It is this pathway that we explore in Chapter 4 by presenting some decision-aiding methods.

4

The Decision: Methods

4.1. Introduction

In Chapter 3, we took the time to discover the current of MCDA (Multi-Criteria Decision-Aiding, see section 2.5) and its approach for characterizing the decision problem as well as its essential notions, i.e. preference, on the one hand, and order, on the other hand. It was therefore possible for us to show that the different types of preference considered only allowed the development of pre-orders, which could not systematize the choice and therefore the decision. Through the central stage of the decision process, this chapter will address the methods proposed in MCDA, what and how they bring the "extra" needed to come closer to order, which enables choice. Coming closer, yes, to greater or lesser degrees, but coming closer only. Deducing a total order would rely, in this formalism, on the principle of stating a number of hypotheses, which cannot be coherent with the vision that MCDA has of decisions and the decision-makers' involvement in the process. Moreover, although the basic idea is still choosing from a comparison (so from an order), an additional principle is specified, that of "synthesis". This synthesis or even combination or aggregation has the aim of bringing together "information on different viewpoints or aspects concerning a set of objects (or alternatives, actions), and choosing one or more objects from among this set" [GRA 05a].

Moreover, although two initially separate schools founded the current of MCDA – the European school and its principle of outranking, on the one

---

For a color version of all the figures in this chapter, see www.iste.co.uk/berrah/decision.zip.

hand, and the American school and its principle of aggregation, on the other hand, the boundaries between these two schools are no longer so impermeable, with some methods having appeared more recently that rely on both principles at once. Its geographical origins had already been partly abandoned. However, mentioning its origins has been important, as it has allowed us to draw out links between the decision-making contexts and the schools chosen to do this, at least for the European school and its eastern heritage (see section 1.2.8). Centered on the formal aspect of the methods, we will therefore distinguish in this chapter not the origin of the school, but its principle, which is outranking for the European school and aggregation for the American school. This chapter will therefore be structured into two main sections that will recall the principles, strengths and limitations of the significant methods of each school, as well as the associated software solutions. Naturally, and in conformity with the spirit of this book, this chapter will not be able to substitute an exhaustive view of the implementation of these methods by their authors or the summaries made in articles and reference works such as [SCH 85; GUI 98; FIG 05a]. It will, however, be original in providing some elements of positioning the methods in relation to one another, in view of industrial practice, illustrated by the problem of choosing an MES, introduced previously, and naturally, of the decision-maker's role when applying these methods.

It is therefore a question of choosing one alternative depending on a number of criteria. The notations proposed in Chapter 3 will be retained in this chapter. The book's care for formalization will be systematized so as to present homogenously the different methods chosen. In this sense, we will complete and propose, according to need, the notations and formulations needed to present the methods.

> To acquire an MES, the information system manager had drawn information from the table of results that compared the alternatives considered (see section 3.4). Having obtained a partial pre-order that took account only of incomparability relationships, they were advised to pursue the analysis in conformity with the method envisaged. Nonetheless, in this more in-depth study, they will consider all six alternatives, $A=\{a_1,a_2,a_3,a_{14},a_5,a_6\}$ introduced previously (see section 3.4.1).

4.2. Outranking

4.2.1. *Principles*

The school of outranking revisits the notion of preference. It specifies the latter in a preference called "outranking" defined as follows:

– an alternative a outranks an alternative b, (aSb), when it is considered to be at least as good as it; in this case, a is strictly preferred to b or is indifferent to it, i.e.: $r(a) \leq r(b)$ where $r(a)$ and $r(b)$ are the ranks of alternatives a and b, respectively (see section 3.4.3);

– an alternative a does not outrank an alternative b, $(a\neg Sb)$, when b is strictly preferred to a, i.e.: $r(a) > (b)$.

The outranking relationship is an order relation that thus presents the following properties:

– if $(aSb) \wedge (bSa)$, then there is indifference between a and b;

– if $(aSb) \wedge (b\neg Sa)$, then a is strictly preferred to b;

– if $(a\neg Sb) \wedge (b\neg Sa)$, then there is incomparability between a and b.

Moreover, the outranking relationship separates the criteria into two ranks:

– criteria in concordance with the proposition (aSb);

– criteria in discordance with the proposition (aSb).

If the number of criteria "in concordance" with the proposition (aSb) is higher than (or at least the same as) the number of criteria "in discordance" with this proposition, then this will be validated. In this opposing case, the proposition $(a\neg Sb)$ will be validated. Although this intuitive formulation may appear to be very clear, it does not provide more variants when it involves a comparison that goes beyond two alternatives, as for the manner of defining the relationship between the criteria in concordance and those in discordance. The notions of concordance and discordance therefore become quantified and nuanced.

The power of outranking lies in the simple explanation of how the order is obtained from the "for" (the criteria in concordance) and "against" (the criteria in discordance). The choice will therefore result from the "synthesis" of both sets of information. In this, the principle still relies on a comparison, two by two, of the alternatives considered (see section 3.3). It is nevertheless asked of the decision-maker to give, in addition, the "rules and conventions" that make it possible to establish the concordance and discordance information, on the one hand, and to draw a conclusion from this, on the other hand (see section 3.5.2). The first limitation of this approach is, as stated, that it does not establish a total order, situations of indifference remain, as well as all or some incomparability situations. Its second limitation is that it remains reserved for an informed public, with the intellectual curiosity to leave behind the quantification methods popular among engineers. In conformity with the principle of MCDA, two different decision-makers faced with the same alternatives and the same criteria for choosing can reach different choices. Is this a good thing? Is it a bad thing? Still, although this principle of outranking is shared by the methods of its school, variants will remain, depending on the method. Depending on the method considered, the decision-maker will have the freedom to choose the one that suits them.

Indeed, MCDA proposes several methods based on the notion of outranking. The interested reader will be able to learn more about the methods of outranking in [FIG 05a; MAR 05]. The oldest method is Condorcet's, removed from its purpose of analyzing election methods [NUR 10]. More recently, in the 1960s, the ELECTRE (*ELimination Et Choix Traduisant la REalité* – Elimination and Choice Translating Reality) family of methods was gradually formed [ROY 91; ROY 93], in line with refinements of its base algorithm[1]. In the 1980s, the more sophisticated family of PROMETHEE (Preference Ranking Organization Method for Enrichment Evaluation) methods was formed [BRA 02; BRA 05], also on the basis of refinement of its algorithm and its graphical tools[2].

---

1 The ELECTRE I method appeared, then evolved into ELECTRE II and finally into ELECTRE III, before being supplemented by ELECTRE Tri, specific to classification problems.
2 The PROMETHEE I method appeared, and was then supplemented by the PROMETHEE II method and by GAIA, its dedicated graphical representation tool.

We have chosen, in the context of this book, to present Condorcet's method for its simplicity and the ELECTRE III method for its frequency of use.

4.2.2. Condorcet's method

The origins of the school of outranking go back to Condorcet's work on the notion of majority, applied in the context of choosing an assembly representative (see section 3.3). Condorcet's method enabled problems of choice and ranking to be addressed (see section 3.4.3) and, today, it remains confidential in the domain of industry.

An alternative a outranks an alternative b, (aSb), in Condorcet's sense, if there are at least as many criteria in concordance with this proposition as in discordance with the latter. Following this principle, three cases may be encountered:

– parity, for which there are as many criteria in concordance as criteria in discordance with the proposition (aSb). In this case, $(aSb) \wedge (bSa)$, and there is indifference between a and b;

– the simple majority, for which there are more criteria in concordance than criteria in discordance with the proposition (aSb). In this case, $(aSb) \wedge (b\neg Sa)$, and a is strictly preferred to b;

– the minority, for which there are fewer criteria in concordance than criteria in discordance with the proposition (aSb). In this case, $(a\neg Sb) \wedge (bSa)$ and b is strictly preferred to a.

It should be noted that the case $(a\neg Sb) \wedge (b\neg Sa)$ cannot be met, since it is impossible to have a simple majority of criteria for both of the alternatives. Condorcet's method eliminates cases of incomparability and maintains situations of *ex aequo*. The alternatives therefore form a complete pre-order.

On the basis of the simple majority principle of Condorcet's method. The information system manager will be able to compare the alternatives of the set $A=\{a_1,a_2,a_3,a_{14},a_5,a_6\}$ described in Table 3.4 to reach a conclusion on the outranking relationships, summarized in the outranking matrix in Table 4.1. In this instance, they will be able to observe (bold entry in Table 4.1) that:

– a_1 is at least as good as a_2 according to the criteria *Acquisition_price, Technical_feasibility_of_implementation* and *Collaboration_level*;

– a_2 is at least as good as a_1 according to the criteria *Response_to_specifications* and *Associated_service*.

Three criteria will be concordant with the proposition $a_1 S a_2$, two criteria will be discordant with this proposition. There will therefore be more criteria in concordance than criteria in discordance, in consequence: $a_1 S a_2$.

Similarly:

– $a_2 \neg S a_1$, two criteria will be concordant with the proposition $a_2 S a_1$ and three criteria will be discordant with this proposition;

– $a_1 S a_3$, three criteria will be concordant with the proposition $a_1 S a_3$ and two criteria will be discordant with this proposition;

– $a_3 S a_1$, three criteria will be concordant with the proposition $a_3 S a_1$ and two criteria will be discordant with this proposition.

In the end, the information system manager will be able to establish strict preference and indifference relationships between all the alternatives. For example:

– $a_1 S a_2 \wedge a_2 \neg S a_1$, consequently, a_1 is strictly preferred to a_2;

– $a_1 S a_3 \wedge a_3 S a_1$, consequently, a_1 and a_3 are indifferent;

– $a_1 S a_4 \wedge a_4 S a_1$, consequently, a_1 and a_4 are indifferent;

–etc.

Figure 4.1 presents the Hasse diagram corresponding to the outranking relationship obtained.

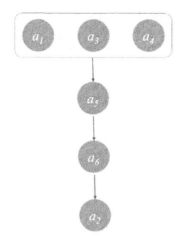

Figure 4.1. *Hasse diagram in Condorcet's method*

The information system manager will therefore be able to eliminate a_2, a_5 and a_6. This is progress compared to analyzing preferences that would identify these six alternatives as incomparable (see section 3.6). Nevertheless, they will still be faced with three alternatives that are indifferent compared to one another.

Alternative	a_1	a_2	a_3	a_4	a_5	a_6
a_1		**3/2** $\mathbf{a_1 S a_2}$	3/3 $a_1 S a_3$	3/3 $a_1 S a_4$	4/2 $a_1 S a_5$	3/2 $a_1 S a_6$
a_2	2/3 $a_2 \neg S a_1$		1/4 $a_2 \neg S a_3$	2/4 $a_2 \neg S a_4$	2/3 $a_2 \neg S a_5$	1/4 $a_2 \neg S a_6$
a_3	3/3 $a_3 S a_1$	4/1 $a_3 S a_2$		3/3 $a_3 S a_4$	4/2 $a_3 S a_5$	5/0 $a_3 S a_6$
a_4	3/3 $a_4 S a_1$	4/2 $a_4 S a_2$	3/3 $a_4 S a_3$		3/2 $a_4 S a_5$	4/2 $a_4 S a_6$
a_5	2/4 $a_5 \neg S a_1$	3/2 $a_5 S a_2$	2/4 $a_5 \neg S a_3$	2/3 $a_5 \neg S a_4$		4/2 $a_5 S a_6$
a_6	2/3 $a_6 \neg S a_1$	4/1 $a_6 S a_2$	2/4 $a_6 \neg S a_3$	0/5 $a_6 \neg S a_4$	2/4 $a_6 \neg S a_5$	

Table 4.1. *Outranking matrix in Condorcet's method*

The strong aspect of Condorcet's method lies in its simplicity and the systematic way in which it is applied.

Indeed, the decision-maker's role in applying this method will remain lesser and will be centered on defining their problem exactly. Two different decision-makers will thus have the same choices for the same problem. On the contrary, the main limitation of this method lies in the fact that it allows cases of indifference to subsist when the concordant criteria and the discordant criteria are the same in number, even though variants allowing criteria to be weighted have been proposed [NUR 10]. Moreover, the Condorcet method remains time-consuming to implement because of the way it processes information, which requires carrying out a potentially large number of comparisons, $m \times (m-1)$ for m alternatives. If no software solution is linked to the method, the decision-maker controlling the spreadsheets will be able to systematize the processing by developing their own calculating sheet or using online services dedicated to multi-criteria decision-aiding (Decision Deck[3], for example).

4.2.3. *The ELECTRE III method*

4.2.3.1. *General remarks*

As the first methods claiming to belong to the school of outranking, ELECTRE methods have made it possible to tackle decision problems other than making choices (see section 3.4.3). They gradually introduced information elements, ("rules and conventions", see section 3.5.1), making it possible to refine the rankings obtained. Thus, ELECTRE I considers the notion of a "threshold" to give nuance to the discordance [ROY 68a]. ELECTRE II adds to it the notion of a "veto" to reinforce the discordance in the extreme. For its part, ELECTRE III, since it considers ranking problems in addition, would add more sophistication to addressing the notion of a threshold so as to take into account the imperfection of the data and would introduce the notion of weight for the criteria. ELECTRE III would naturally be substituted by two other methods that it would generalize. We note moreover the ELECTRE Tri method, the development of which was proposed in parallel with that of the ELECTRE III method. ELECTRE Tri tackles sorting problems and, to define the classes, adds the notion of "reference"

---

3 http://www.decision-deck.org/project/.

alternatives to the previous notions (see section 3.4.3) [YU 92]. Despite the training required to use them, ELECTRE methods are echoed in some ways in a number of domains including the industrial domain [GOV 16][4].

An alternative a outranks an alternative b, (aSb), in the sense of ELECTRE III if the criteria in concordance obtained a qualified majority, i.e. a more substantial majority than the simple majority, without there being any "strong" discordance between them. The principle of ELECTRE would rely precisely on quantifying this qualification as "strong". To do this, the method refines the notions of concordance and discordance by introducing thresholds, making it possible to qualify the latter (according to each criterion). It then proposes a synthesis mechanism allowing the outranking. Figure 4.2 presents a synopsis positioning all the information used by the method and the stages of developing the choice.

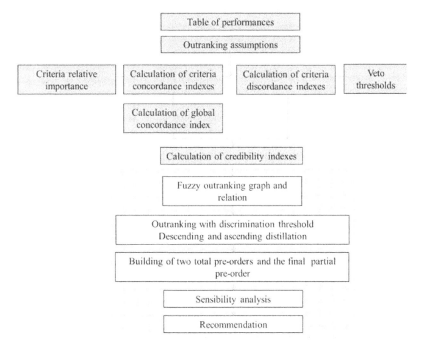

Figure 4.2. *Synopsis of the ELECTRE III method [MAY 94]*

---

4 In our work, the ELECTRE III method has been applied to industrial governance [BER 15], for improving the performance of SMEs (small and medium-sized enterprises) [CLI 13] or even for choosing locations for installing photovoltaic panels in urban environments [THE 20].

In coherence with the aim of this chapter, only the stages requiring the intervention or comprehension of the decision-maker will be detailed, since the systematic stages have broadly been presented by the method's authors [FIG 05b] and in some cases reprised by other researchers (see [MOU 11], for example, for sensitivity studies). Hence, we will address:

– defining the said outranking hypotheses;

– developing concordance and discordance indexes, and synthesizing them as well as the vetoes.

Like all MCDA methods, ELECTRE III is applied to previously posed decision problems, for which the performance table is available (see section 3.4.3). However, since the principle of ELECTRE III is to link thresholds to the criteria, the latter should hence be quantified and not ranked [MAY 94].

To choose their MES, the information service manager had defined a table of results (see section 3.4.2). In the case of applying ELECTRE III, the first two criteria had been obtained by an identify function and would take the value of the corresponding attributes (Table 3.2). For the three other criteria, the information system manager would opt to quantify them, depending on their expertise. For the sake of harmony, the whole values are defined on a scale of 0–10, 10 expressing the information system manager's total satisfaction and 0 no satisfaction. The new table of results is given in Table 4.2.

Criterion / Alternative	g_1	g_2	g_3	g_4	g_5
a_1	9,280	4	4	7	4
a_2	34,000	5	8	3	6
a_3	13,660	7	6	5	7
a_4	10,890	6	6	7	6
a_5	15,000	4	7	4	6
a_6	16,500	6	5	4	4

Table 4.2. *Table of results for choosing an MES in ELECTRE III*

4.2.3.2. Defining outranking hypotheses

ELECTRE III asks the decision-maker to define their "hypotheses" by linking to each g_i criterion two different types of threshold, the indifference threshold, q_i, and the preference threshold, p_i. These thresholds are defined by the generally positive values, stated on the definition domain, G^i of g_i. Defining these thresholds allows the subsequent establishment of preference relationships between alternatives. Thus, for a pair of alternatives (a, b) whose g_i values are written as c_{ia} and c_{ib} (see section 3.4.2):

— a and b are indifferent in the sense of g_i if the value taken by a is higher than that taken by b by at most the value of q_i, or if the value taken by b is higher than that taken by a by at least the value of q_i: $|c_{ia} - c_{ib}| \leq q_i$, $|c_{ib} - c_{ia}| \leq q_i$;

— a is strictly preferred to b in the sense of g_i if the value taken by a is higher than that taken by b by at least the value of p_i: $(c_{ia} - c_{ib}) \geq p_i$;

— a is weakly preferred to b in the sense of g_i if the value taken by a is strictly higher than that taken by b by at least the value of q_i while still remaining strictly lower than the value of p_i: $p_i : q_i < (c_{ia} - c_{ib}) < p_i$.

For any pair of alternatives, the comparison is thus established, for each criterion, according to the two thresholds, as illustrated in Figure 4.3.

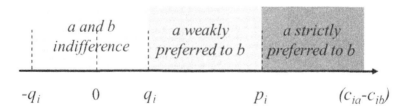

Figure 4.3. *Preferences and thresholds q_i and p_i for the g_i criterion*

We note that it is possible to give nuance to strict preference (see section 3.5.2) using the intermediate notion of weak preference.

By considering the *Acquisition_price* criterion for choosing an MES, the information system manager will be able to:

– be indifferent between two MES up to a price difference of 2,000 €: $q_1 = 2,000$ €;

– prefer strictly one MES that is cheaper than another by 4,000 €: $p_1 = 4,000$ €.

A definition of all the indifference and preference thresholds for the five criteria considered is given in Table 4.3.

Criterion Threshold	g_1	g_2	g_3	g_4	g_5
q_i	2,000 €	0	0	0	0
p_i	4,000 €	1	1	1	2

Table 4.3. *Indifference and preference thresholds in ELECTRE III*

From the table of results (Table 4.2), the information system manager will be able to show that, on the basis of the *Acquisition_price*:

– a_1 is strictly preferred to a_3, (13,660–9,280) ≥ 4,000;

– a_4 is weakly preferred to a_3, 2,000<(13,660–9,280)<4,000;

– a_1 and a_4 are indifferent, |10,890–9,280|≤2,000.

4.2.3.3. *Developing concordance and discordance indexes and synthesizing them*

From the indifference and preference thresholds, it is a question, in the ELECTRE III procedure, of quantifying concordance (or acquiescence) and discordance, this time by means of two types of index:

– the concordance indexes in favor of outranking a in relation to b when the q_i threshold is exceeded; these indexes are both calculated for each criterion: $c_i(aSb)$ and globally: $c(aSb)$;

– the discordance indexes against outranking the alternative a compared to b when the p_i threshold is exceeded, calculated only for each criterion: $d_i(aSb)$.

The concordance and discordance indexes are defined over the interval $[0,1]$, where the value 1 signifies, depending on the case, total concordance or total discordance, in favor of the outranking proposition considered, and the value 0 signifies, depending on the case, total non-concordance or total non-discordance. All values are possible between these bounds.

4.2.3.3.1. Concordance index

$c_i(aSb)$ is calculated for each g_i as follows:

$$c_i(aSb) = \frac{p_i - \min(c_{ia} - c_{ib}, p_i)}{p_i - \min(c_{ia} - c_{ib}, q_i)}$$

For its part, $c(aSb)$, which quantifies the degree of concordance with the proposition (aSb), is obtained from a synthesis of $c_i(aSb)$. This index considers the relative importance of each g_i, by attributing to it a weight w_i. The synthesis is achieved using a weighted average. Calculating $c(aSb)$ then takes the following form:

$$c(aSb) = \frac{\sum_{i=1}^{n} w_i \times c_i(aSb)}{\sum_{i=1}^{n} w_i}$$

The result proposed by this index is a ratio that positions the (weighted) number of the criteria in "concordance" compared to the total (weighted) number of criteria. Therefore, when $c(aSb)=1$, all the criteria will be concordant in favor of the proposition (aSb), i.e. a will be indifferent or preferred to b according to all the criteria. On the contrary, when $c(aSb) = 0$, no criterion will be concordant in favor of the proposition (aSb), i.e. a will be not preferred to b according to all the criteria.

The results of calculating the concordance indexes relating to the choice of MES are presented in Table 4.4. The information system manager will be able to attribute an initial weight to each criterion depending on its importance in the choice: $w_1 = 5, w_2 = 7, w_3 = 3, w_4 = 3, w_5 = 2$. As a form of illustration, the following examples detail the indexes corresponding to the bold entries in Table 4.4:

–the value $c(a_1 Sa_2) = 0.42$ means that a_1 is indifferent or preferred to a_2 to 42%;

–the value $c(a_6 Sa_3) = 0.22$ means that a_6 is indifferent or preferred to a_3 to 22%;

–the value $c(a_4 Sa_2) = 0.84$ means that a_4 is indifferent or preferred to a_2 to 84%;

–the value $c(a_3 Sa_6) = 1$ means that a_3 is indifferent or preferred to a_6 according to all the criteria.

	a_1	a_2	a_3	a_4	a_5	a_6
a_1	1.00	**0.42**	0.42	0.42	0.84	0.42
a_2	0.58	1.00	0.16	0.26	0.58	0.26
a_3	0.72	0.84	1.00	0.80	0.84	**1.00**
a_4	0.86	**0.84**	0.47	1.00	0.74	1.00
a_5	0.68	0.42	0.46	0.31	1.00	0.58
a_6	0.68	0.84	**0.22**	0.53	0.82	1.00

Table 4.4. *Concordance indexes for choosing an MES in ELECTRE III*

4.2.3.3.2. Discordance index

Calculating the discordance index integrates a specific notion, the veto, which expresses an impossibility of the preference relationship according to a given criterion. The veto is conveyed by a threshold written as ω_i according to the g_i criterion. Exceeding this threshold indicates that it is sensible to deny the proposition (aSb) any credibility. This threshold expresses a ban in the comparison mechanism. If the value taken by a is higher than that taken by b by the value of ω_i or more, b cannot in any case be preferred to a whatever the values taken according to the other criteria. In this case, it is said that there is a total discordance of the proposition (aSb). The veto does not therefore "give nuance" to the preference in view of each criterion as such, it conveys the categoric notion of "non-preference" in view of the globality of the criteria. Implementing the veto remains optional, highly dependent on the decision problem and on the decision-maker. Indeed, although there are some cases where any decision-maker would have the same position in the question of a veto, in a dangerous situation for example, there will be others where this is not the case. A decision-maker could consider it prudent to define a veto according to such a criterion while another might see it as an excess of concern, for example.

When the decision-maker wishes to express a ban for g_i, they therefore define a threshold veto ω_i, the value of which is generally positive and stated on the definition domain G^i. Therefore, for a pair of alternatives (a,b), b cannot be preferred to a if the value taken by a is higher than that taken by b by at least $\omega_i : (c_{ia} - c_{ib}) \geq \omega_i$.

> By considering the *Acquisition_price* criterion when choosing an MES, the information system manager cannot prefer one MES to another if the price difference is greater than 10,000 €:ω_1=10,000 €.
>
> In the same way, the information system manager may feel the need to define a veto threshold for g_2 and g_3 and not express one for g_4 and g_5. Thus, according to the *Response_to_specifications*, a significant difference in the number of functions offered by the software, estimated at four, could be felt to be prohibitive. Similarly, according to the *Technical_feasibility_of_implementation* criterion defined by satisfaction levels on a scale of 0–10, a difference in satisfaction level of four could also be experienced as prohibitive (Table 4.2). Concerning the *Associated_service* and *Collaboration_level* criteria, the information system manager may feel there is no prohibition in comparisons according to these criteria and may not define veto thresholds.

A definition of all the veto thresholds for the three criteria considered is given in Table 4.5. From the results table (Table 4.2), the information system manager will therefore be able to see that according to the *Acquisition_price* criterion, the only case of veto would involve the proposition $a_2 S a_1$, a_2 is not preferred to a_1: (34,000 € − 9,280 €) ≥ 10,000 €

Criterion Threshold	g_1	g_2	g_3	g_4	g_5
ω_i	10,000 €	4	4		

Table 4.5. *Veto thresholds in ELECTRE III*

For a given criterion, the discordance index expresses the degree of discordance in the decision-maker's preference to the proposition (aSb). All the comparisons that are partially opposed (strict preferences) or totally opposed (vetoes) to the proposition (aSb) contribute to this index defined on the interval $[0,1]$. The closer this degree to value 1, as is the case when there is a veto, the more the proposition (aSb) is discordant. On the contrary, the closer this degree to value 0, the less discordant it is. Just as for the concordance, the decision-maker will be allowed, here too, to validate the preferences they have formulated.

Calculating the discordance index therefore depends on comparing $(c_{ia} - c_{ib})$ and the p_i and ω_i thresholds. It takes the following form:

$$d_i(aSb) = \begin{cases} 1 \text{ if } (c_{ia} - c_{ib}) \geq \omega_i \\ 0 \text{ if } (c_{ia} - c_{ib}) < p_i \\ \dfrac{(c_{ia} - c_{ib}) - p_i}{\omega_i - p_i} \text{ in the other cases} \end{cases}$$

Over the interval $[0,1]$ all nuances are possible, the closer the value to 1, the greater the discordance, the closer the discordance to 0, the weaker it will be. Unlike the concordance indexes, which quantify a semantics of proportion, the discordance indexes will convey a semantics of removal or distance from the proposition to prefer one alternative to another. Because of this, it will not be possible, even though once again intuition may push us to do so, to consider a form of opposition and correlation between concordance and discordance.

The results of calculating the discordance indexes relating to the choice of MES are presented in Table 4.6. For the sake of readability, only an extract on comparing a_1 with a_2 and a_3 according to the five criteria is provided. As an illustration, the following examples detail the indexes according to g_1, g_2, g_3 corresponding to the bold entries in Table 4.6:

– the value $d_3(a_1Sa_2) = 1.00$ means that preferring a_2 compared to a_1 according to g_3 reaches or exceeds the veto threshold ω_3: $8 - 4 \geq 4$;

– the value $d_1(a_1Sa_2) = 0.00$ means that there is no strict preference of a_2 compared to a_1 according to g_1: 9,280 € –34,000 € <10,000 €;

– the value $d_2(a_1Sa_3) = 0.67$ means that preferring a_3 compared to a_1 according to g_2 exceeds p_2 and more precisely: $d_2(a_1Sa_3) = \dfrac{7-4-1}{4-1} = 0.67$;

– the value $d_3(a_1Sa_3) = 0.33$ means that preferring a_3 compared to a_1 according to g_3 exceeds p_3 and more precisely: $d_3(a_1Sa_3) = \dfrac{6-4-1}{4-1} = 0.33$.

Criterion		Alternative a_2	Alternative a_3
g_1		$d_1(a_1Sa_2) = 0.00$	0.00
g_2	Alternative a_1	0.00	**0.67**
g_3		**1.00**	**0.33**
g_4		0.00	0.00
g_5		0.00	0.00

Table 4.6. *Discordance indexes for pairs (a_1, a_2) and (a_1, a_3) in ELECTRE III*

4.2.3.3.3. Synthesis: degree of credibility

ELECTRE III proposes a mechanism for synthesizing the concordance index $c(aSb)$ and the discordances indexes $d_i(aSb)$ by means of a new index called a degree of credibility, and written as $\rho(aSb)$. The degree of credibility, as its name indicates, quantifies the credibility of the outranking hypothesis. This credibility is the result of highlighting the concordance and associated discordances. The method's authors relate the credibility to a majority of concordant criteria without great discordances.

The degree of credibility is defined over the interval $[0,1]$, where the value 1 means total credibility in favor of the outranking proposition considered, and the value of 0 means total non-credibility. All values are possible between these bounds.

The degree of credibility is the result of a sort of acquiescence to the proposition (aSb) (value of the concordance index) diminished by a kind of refusal of this proposition (value of the discordance index), in conformity with the following formula:

$$\rho(aSb) = c(aSb) \times \prod_{\{i \in I : d_i(aSb) > c(aSb)\}} \frac{1 - d_i(aSb)}{1 - c(aSb)}$$

Two particular cases can be distinguished as follows:

– $\rho(aSb) = c(aSb)$ when $d_i(aSb) \leq c_i(aSb)$ $\forall i \in (1,...,n)$;
– $\rho(aSb) = 0$ when $\exists\ d_i(aSb) = 1$, $i \in (1,...,n)$.

From confirmation (high degree of credibility) of the proposition (aSb) or its non-confirmation (low degree of credibility) $(a \neg Sb)$ ELECTRE III develops an order between the alternatives in the same spirit as the Condorcet method. Concretely, a high degree of credibility of the proposition (aSb) and a low degree of credibility of the proposition $(b \neg Sa)$ lead to the conclusion that $r(a) < r(b)$.

Finally, after having filled in the results table as well as the value of the weight of the criteria and calculated the different indexes, the information system manager will be able to obtain a partial pre-order, as shown in Figure 4.4.

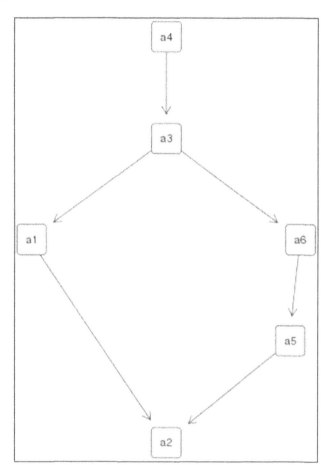

Figure 4.4. *Hasse diagram in the ELECTRE III method*

It appears that:

$-a_4$ is strictly preferred to all other alternatives;

$-a_3$ is strictly preferred to all other alternatives except a_4;

$-a_1$ and a_5 are incomparable;

–all the alternatives are preferred to a_2.

This is progress compared to analyzing preferences that would identify these six alternatives as incomparable (see section 3.6). By "chance", the incomparabilities will not have prevented a_4 from being identified as the alternative outranking all the others. It will be noted, however, that the order obtained using the ELECTRE III method is different to that given by the Condorcet method. In particular, a_4 is the only alternative in rank 1, and the order of a_5 and a_6 has been reversed. The information system manager will thus be able to confirm that the order of the alternatives generally depends on the MCDA method chosen, an observation they could confirm as they continue to experiment with different methods.

Nonetheless, the information system manager would, thanks to information processing, reach the goal of their choice.

4.2.3.4. Strengths and limitations

The decision support provided by ELECTRE III thus consists of recommending one or potentially several alternatives. The decision-maker nevertheless has a dashboard formed of three families of indicators (the different concordance, discordance and credibility indexes, developed for each pair of alternatives). This gamble was successful, as shown by the immediate and easy use of this dashboard. However, the information the decision-maker has is so intuitive, complete and close to their reasoning that, without any doubt, only information and additional processing (rules, trends) could allow a better appropriation of it.

The strength of the ELECTRE III method is simple reasoning based on "fors" and "againsts". This reasoning, however, requires the decision-maker to give nuance to their preferences (indifference, preference thresholds), to express their non-preferences (veto thresholds) and to reflect on their priorities (the weight of the criteria). The method does not remain any less complex to master. Its main limitation remains the establishment only of partial pre-orders. It will essentially be this limitation that will have given weight to the school of aggregation, requiring a total order in almost all cases.

The first solution associated with ELECTRE III was developed by the method's authors in the 1980s. Today, it is possible to use online services dedicated to multi-criteria decision-aiding (also Decision Deck[5]).

4.3. Aggregation

4.3.1. *Principles*

The school of aggregation revisits the notion of preference by linking it to a numerical quantification, i.e. values defined according to cardinal scales[6] [STE 46]. In the particular case, where these values provide information on the decision-maker's satisfaction with regard to the alternatives, they are called utilities (see section 3.3). It is called a "marginal" utility when a sole criterion g_i is considered to express this satisfaction. It will be called a global utility when all the criteria are considered. In this, the school of aggregation is supported by MAUT (Multi-Attribute Utility Theory) and MAVT (Multi-Attribute Value Theory) [KEE 76; DYE 05].

Marginal utilities, traditionally written as $u(v_i(a))$ ($i = 1,...,n$), more simply $u_i(a)$, are obtained from the v_i attributes. When they are normalized, as in the vast majority of cases, they are merely a re-naming of the g_i criteria (see section 3.4.2).

(Global) utility written as $U(a)$ is the result of aggregating $u_i(a)$ by an aggregation function generally written as F:

---

5 http://www.decision-deck.org/project/.
6 A scale makes it possible to describe an attribute by a quantity, whether qualitative or quantitative. When the quantities are defined by a numerical value, it is called a cardinal scale. When the quantities are bounded by two distinct numerical values, it is called a pre-cardinal scale. [BAN 97]. Quantitative scales are distinguished according to operations authorized on the quantities. So for two utilities $u(a)$ and $u(b)$, it is called an ordinal scale when the comparison ($u(a) > u(b)$, for example) has a meaning, an interval scale when the difference $u(a) - u(b)$ has a meaning and a ratio scale when the $\frac{u(a)}{u(b)}$ ratio has a meaning.

$$F: U^1 \times ... \times U^n$$
$$(u_1(a),...,u_n(a)) \to U(a) = F(u_1(a),...,u_n(a))$$

where U^i is the set of values taken by $u_i(a)$, and U is the set of values taken by $U(a)$.

The order between the alternatives considered will therefore conform to the order between their (global) respective utilities:

– a is strictly preferred to b if $U(a) > U(b)$;

– a and b are indifferent if $U(a) = U(b)$.

Since the comparison between $U(a)$ and $U(b)$ is always an equality or an inequality[7], the incomparability from which outranking methods suffer can no longer happen using this approach.

The school of aggregation, which takes its name from the aggregation function of the marginal utilities, proposes several methods [DIA 92]. These will differ depending on how the marginal utilities are defined $u_i(a)$ and/or how the aggregation operation F is carried out. Aggregation is generally achieved by means of a sum or average operator, the most widely used operator remains the WAM (Weighted Arithmetic Mean), the weights relating to the different criteria. The interested reader will be able to discover more aggregation methods in [GUI 98; FIG 05a; MAR 05].

The strength of the aggregation approach lies in a quantification that is easy to interpret. It can also seem potentially reassuring because of its formalization principle. By virtue of their belief in the principle, the decision-maker would be ready for any cognitive attempts to determine the

---

[7] We note that the utilities can be defined in the form of an interval and not of a single value. They will be bounded in this case by a min value $U_{\min}(a)$ and a max value $U_{\max}(a)$: $U_{\min}(a) \leq U(a) \leq U_{\max}(a)$. Given the uncertainty conveyed by the estimation of the utilities [GRE 08], it is possible to encounter situations of incomparability called "possible" preferences.

value of the criteria and the parameters of the aggregation function. However, the limitation of this school will also be, in this principle, a reasoned determination of these values requiring mathematical bases that are often unknown to the decision-maker, and indeed neglected by the authors of some methods themselves. Not all quantifications are good and the information processing carried out should respect some requirements so that the information obtained can have meaning for the decision. This being the case, in conformity with the principle of MCDA and as for outranking, two different decision-makers, faced with the same alternatives and the same criteria for choice, can reach different quantifications and so different choices. Given this, it will also remain for the decision-maker to choose their aggregation method, while remaining aware of the hypotheses to which they should assent to quantify their decision, and of the criticisms that will inevitably be made of them.

The oldest aggregation method is the Borda count, "diverted" from its purpose for elections, just like the Condorcet method for outranking. It would nevertheless be the much more recent work of Fishburn on additive utilities [FIS 70; FIS 81] that would allow the general framework of aggregation to be put in place. The basic hypothesis behind this framework is that the "whole", the global utility, is the sum of the "parts", the marginal utilities. In this matter, a number of conditions should be respected when defining the marginal utilities and choosing the aggregation operator KRA 71]. From this hypothesis, aside from the MAUT and MAVT theories already cited, the school of aggregation saw the success of many new methods which the reader will be able to consult in [BRA 10; MAR 15]. Among these methods, we can cite the SMART (Simple Multi-Attribute Rating Technique) [EDW 71] method, the very widely used AHP (Analytic Hierarchy Process) method [SAA 77], the UTA (additive utilities) method [JAC 82], the TOPSIS (Technique for Order of Preference by Similarity to Ideal Solution) method [HWA 94] and finally the MACBETH (Measuring Attractiveness by a Categorical Based Evaluation TecHnique) [BAN 97] method.

We choose, in the context of this book, to present the Borda count method for its simplicity, the AHP method for its reputation and the MACBETH method for its mathematical coherence and its total respect for the principles of MCDA.

4.3.2. *The Borda count method*

With its principle of adding the ranks of different candidates in an election to choose the one with the least high global score (see section 3.3), the *chevalier* de Borda provided the inspiration for the principle of aggregating for decisions, even though this did not respect the conditions, the sum, that should be met to use an additive aggregation operator, and the notions of utilities were not mentioned. This fairly intuitive method was often used to tackle choosing and ranking problems (see section 3.4.3). This method was very broadly applied in the industrial domain without its origin necessarily being known, with engineers systematically ranking alternatives according to criteria to then add up the rankings obtained to create a global score.

Alternative a is strictly preferred (respectively indifferent) to alternative b in Borda's sense if the sum of the ranks of the criteria associated with a is lower (respectively equal) to the sum of the ranks of the criteria associated with b. The Borda method eliminates the incomparabilities and maintains the situation *ex aequo*. The alternatives therefore form a complete pre-order.

Following the principle of the Borda method, the IT system manager will be able to compare the alternatives of the set $A=\{a_1,a_2,a_3,a_4,a_5,a_6\}$ to conclude on the preference and indifference relationships, summarized in the rank matrix in Table 4.7. In this case, they might, for example, observe (bold entries) that:

$$-\sum_{i=1}^{n} r_i(a_1) = 1+5+6+1+5 = 18\,;$$

$$-\sum_{i=1}^{n} r_i(a_2) = 6+4+1+6+2 = 19\,.$$

It appears that a_4 with $\sum_{i=1}^{n} r_i(a_4) = 10$ presents the lowest sum of the ranks and that a_3 follows it with $\sum_{i=1}^{n} r_i(a_3) = 11$. It is therefore strictly preferred to all the others. Figure 4.5 presents the Hasse diagram corresponding to the order obtained.

Rank / Alternative	r_1	r_2	r_3	r_4	r_5	$\sum_{i=1}^{n} r_i(a_j)$
a_1	1	5	6	1	5	*18*
a_2	6	4	1	6	2	*19*
a_3	3	1	3	3	1	*11*
a_4	2	2	3	1	2	*10*
a_5	4	5	2	4	2	*17*
a_6	5	2	5	4	5	*21*

Table 4.7. *Ranks for the Borda method*

Figure 4.5. *Hasse diagram in the Borda count method*

With this method, the information system manager will even be able to establish a total order on the alternatives by "chance". Their choice problem will very simply be solved in this way.

The strength of the Borda count method lies in its simplicity. Its main limitations arise from its scientific foundations. In fact, its postulate of generating ranks from the rank sums is questionable. As such, its main detractor was Condorcet, who was the first to state that it was not right to consider the sum of the ranks as a rank itself, a statement that was later confirmed by measurement theory [KRA 71]. Moreover, its principle of developing orders without explicit comparison contradicted the principle of independence (see section 3.3), which stipulates that the order between two alternatives should depend only on comparing them [ARR 62]. Nevertheless, what can be confusing is that this approach does not clash with the practices of decision-makers, who are not often surprised by the results given and who would even tend to identify their reasoning, where there are ranks, with its principle. If no software solution is associated with the method, the decision-maker mastering the spreadsheets will be able to systematize the processes by developing their own spreadsheet.

4.3.3. The AHP method

4.3.3.1. General remarks

Among the first to use aggregation for multi-criteria decisions, the AHP method addresses choice and ranking problems (see section 3.4.3). Proposed in the 1970s by Saaty [SAA 77], it became almost a standard to the point where many uninformed users identify the school of aggregation, indeed MCDA with the method. In its principle, the AHP method does not resonate entirely with the stages advised by MCDA (see section 3.7), to the extent where the alternatives are compared two by two (pairwise comparison) using a mechanism that dispenses with knowledge, in the strict sense, of the criteria. These criteria are defined by their hierarchical relationship as well as by their mutual weighting. In this sense, it contradicts the definition of a criterion while greatly simplifying the decision-maker's reflection, and its application often gives similar results to those from other methods. These aspects certainly bring elements of explanation to the fact that it is the MCDA approach that is by far the most used, in all domains, especially industry [SIP 10; ISH 11].

To specify the notion of preference, the AHP method introduces nuance via the relative magnitudes of the alternatives compared to one another. In its principle, a relative magnitude reflects the intensity of the preference the decision-maker feels between two alternatives and is expressed linguistically. Moreover, the notion of "priority" serves to quantify the preferences. The idea of creating a hierarchy of criteria is intended to "decomplexify" the reality of the decision problem, called the goal, by granting the possibility of breaking down each of the initial criteria into a number of sub-criteria that will require understanding of the problem. AHP consequently offers the decision-maker, in an analytical vision of the decision problem, the possibility of focusing on each criterion independently of the others. By sub-problem is thus understood the entirety of all the alternatives, the criterion considered, as well as the sub-set of possible sub-criteria that may be linked to them. For each sub-problem, the mechanism for comparing respectively the alternatives and the sub-criteria will allow the calculation, subject to verification, of priorities (alternatives), on the one hand, and the weights (of the sub-criteria), on the other hand. A "cascade" will at first make it possible to obtain the priority of an alternative according to the criterion considered. The WAM operator is used to aggregate the priorities according to the sub-criteria. Second, the overall priority of each alternative will be calculated. The WAM operator is also used to aggregate the different priorities depending on the criteria. The order between the alternatives will then conform to the respective values of this global priority.

Figure 4.6 presents a synopsis positioning the hierarchy of criteria and the acquisition of relative magnitudes, to reach a calculation of weights and priorities. A stage verifying the coherence of the results obtained is enacted before proceeding to the last aggregation stage. As for the ELECTRE III method, the only stages detailed are those that require the intervention or understanding of the decision-maker. This being the case, the stages defining this goal and the structuring are extensively documented by the author of the AHP method [SAA 84]. Hence, we will address:

– the acquisition of relative magnitudes, the calculation of priorities and weights and verification of their coherence;

– the aggregation of priorities.

In reference books on AHP [SAA 84; SAA 05], the method is often presented with the aid of examples without all the concepts used being systematically formalized. For the sake of harmonization with the remainder of this book and for ease of reading, we will continue in the rest of this section to formalize these concepts on the basis of notations introduced previously, while still attempting to be faithful to the method's information processing.

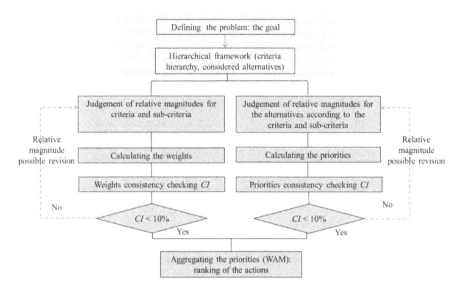

Figure 4.6. *Synopsis of the AHP method*

When implementing the method, it is recommended not to break the criteria down on too many levels (generally, three levels are enough) and into too many sub-criteria (generally three to five sub-criteria are enough). The q sub-criteria relating to g_i can be written as g_{i_k} $(k = 1,...,q)$.

> The information system manager may consider that the goal to be reached is that of choosing an MES. They will be able to envisage breaking down the *Response_to_specifications* criterion. Characterized until now by the number of functions created, this criterion can, in fact, be made more specific. Three sub-criteria can be considered:
>
> – sub-criterion g_{2_1}: *Virtualization* gives information on the MES's ability to simulate production;
>
> – sub-criterion g_{2_2}: *Database* gives information on the MES's ability to interoperate with other applications;

– sub-criterion g_{2_3}: *Interface* evaluates the usability of the software interfaces.

The hierarchy of criteria is presented in Figure 4.7.

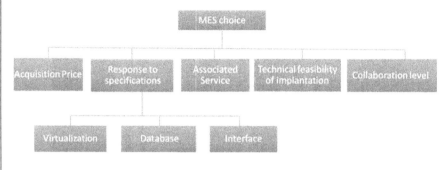

Figure 4.7. *Hierarchy of criteria in AHP*

4.3.3.2. *Acquiring relative magnitudes, calculating priorities*

Concerning the acquisition of relative magnitudes, the method introduces nuance, which it proposes expressing in the form of categories, ranking from equivalence in magnitude to absolute difference in magnitude. It does so based on Miller's principle of scales. Miller considered that humans are able to discern up to seven categories when evaluating a phenomenon [MIL 56]. In this spirit, AHP proposes collecting the relative magnitude $di_i(a_j, a_k)$ of a_j compared to a_k according to g_i, in the form of a ratio of priorities $\dfrac{pr_i(a_j)}{pr_i(a_k)}$, the so-called pairwise comparison where $pr_i(a_j)$, respectively $pr_i(a_k)$, is the priority of a_j (respectively a_k) according to g_i. These ratios are predefined across five categories: "equal magnitude", "moderate (three times) greater magnitude", "strong (five times) greater magnitude", "very strong (seven times) greater magnitude" and "extreme (nine times) greater magnitude". In conformity with the notations adopted in Chapter 3 (see section 3.4.1):

– a_j is of "equal" magnitude to a_k: $di_i(a_j, a_k) = 1$ and $di_i(a_k, a_j) = 1$;

– a_j is of "moderate" relative magnitude compared to a_k: $di_i(a_j, a_k) = 3$ and $di_i(a_k, a_j) = \frac{1}{3}$;

– a_j is of "stronger" magnitude than a_k: $di_i(a_j, a_k) = 5$ and $di_i(a_k, a_j) = \frac{1}{5}$;

– a_j is of "much stronger" magnitude than a_k: $di_i(a_j, a_k) = 7$ and $di_i(a_k, a_j) = \frac{1}{7}$;

– a_j is of "more extreme" magnitude than a_k: $di_i(a_j, a_k) = 9$ and $di_i(a_k, a_j) = \frac{1}{9}$.

Sometimes judged insufficiently explicit, this categorization can distinguish up to nine categories, by introducing intermediate ratio values. The collected $di_i(a_j, a_k)$ form the said pairwise comparison of alternatives:

$$\begin{bmatrix} di_i(a_1, a_1) & \ldots & di_i(a_1, a_j) & \ldots & di_i(a_1, a_m) \\ \ldots & \ldots & \ldots & \ldots & \ldots \\ di_i(a_j, a_1) & \ldots & di_i(a_j, a_j) & \ldots & di_i(a_j, a_m) \\ \ldots & \ldots & \ldots & \ldots & \ldots \\ di_i(a_m, a_1) & \ldots & di_i(a_m, a_j) & \ldots & di_i(a_m, a_m) \end{bmatrix}$$

By considering the *Acquisition_price*, the information system manager will be able to express $di_i(a_j, a_k)$. For example:

– a_1 is of *moderate* relative magnitude compared to: a_3: $di_1(a_1, a_3) = 3$ and $di_1(a_3, a_1) = \frac{1}{3}$;

– a_1 is of *much stronger* magnitude than a_2: $di_1(a_1,a_2)=7$ and $di_1(a_2,a_1)=\frac{1}{7}$;

– a_2 is of *equal* magnitude to a_4: $di_1(a_2,a_4)=1$ and $di_1(a_4,a_2)=1$.

The $di_1(a_j,a_k)$ are assembled in the following matrix:

$$\begin{bmatrix} 1 & 7 & 3 & 2 & 4 & 5 \\ 1/7 & 1 & 1/4 & 1 & 1/3 & 1/2 \\ 1/3 & 4 & 1 & 1 & 2 & 2 \\ 1/2 & 3 & 1 & 1 & 3 & 2 \\ 1/4 & 3 & 1/2 & 1/3 & 1 & 1 \\ 1/5 & 2 & 1/2 & 1/2 & 1 & 1 \end{bmatrix}$$

At this level of knowledge (relative magnitudes of the alternatives in the form of ratios), it is a question of calculating the priorities of the alternatives. These are different to the satisfactions and utilities that we are used to from MCDA and reflect the importance of the alternatives with regard to the goal (the problem). Calculating them involves criteria and sub-criteria at the lowest level of the breakdown. AHP proposes providing priorities in normalized process and proceeds in three stages to do so: normalization, calculation then normalization:

– normalization of the relative magnitudes $di_i{'}(a_j,a_k) = \dfrac{di_i(a_j,a_k)}{\sum_{j=1}^{n} di_i(a_j,a_k)} = 1$,

with $di_i{'}(a_j,a_k)$ being the normalized relative magnitudes;

– calculating the priorities $pr_i(a_j) = \sum_{k=1}^{n} di_i{'}(a_j,a_k)$;

– normalization of the priorities $pr_i{'}(a_j) = \dfrac{pr_i(a_j)}{n}$, with $pr_i{'}(a_j)$ being the normalized priorities.

To obtain the $pr_i{'}(a_i)$ with regard to the *Acquisition_price* criterion, the information system manager will begin by normalizing their $di_1(a_j,a_k)$

and obtaining the following comparison matrix where the calculation results will be rounded to the third decimal:

$$\begin{bmatrix} 0.412 & 0.350 & 0.480 & 0.343 & 0.353 & 0.435 \\ 0.059 & 0.050 & 0.040 & 0.171 & 0.029 & 0.043 \\ 0.137 & 0.200 & 0.160 & 0.171 & 0.176 & 0.174 \\ 0.206 & 0.150 & 0.160 & 0.171 & 0.265 & 0.174 \\ 0.103 & 0.150 & 0.080 & 0.057 & 0.088 & 0.087 \\ 0.082 & 0.100 & 0.080 & 0.086 & 0.088 & 0.087 \end{bmatrix}$$

For example, $di_1(a_1, a_1)$ is normalized by $di_1'(a_1, a_1) = \dfrac{1}{2.426} = 0.412$ given that $\sum_{j=1}^{6} di_1(a_j, a_1) = 1 + \dfrac{1}{7} + \dfrac{1}{3} + \dfrac{1}{2} + \dfrac{1}{4} + \dfrac{1}{5} = 2.426$.

The information system manager will therefore be able to obtain for g_1:

- $pr_1(a_1) = 2.373$;
- $pr_1(a_2) = 0.393$;
- $pr_1(a_3) = 1.019$;
- $pr_1(a_4) = 1.126$;
- $pr_1(a_5) = 0.565$;
- $pr_1(a_6) = 0.523$.

For example, $pr_1(a_1)$ is obtained by the sum of $di_1'(a_1, a_k)$ $k = 1, ..., 6$
$pr_1(a_1) = 0.412 + 0.350 + 0.480 + 0.343 + 0.353 + 0.435 = 2.373$.

In last place, the $pr_1(a_j)$ obtained will be normalized. Therefore:

- $pr_1'(a_1) = 0.395$;
- $pr_1'(a_2) = 0.066$;

$$\begin{vmatrix} -pr_1'(a_3) = 0.170; \\ -pr_1'(a_4) = 0.188; \\ -pr_1'(a_5) = 0.094; \\ -pr_1'(a_6) = 0.087. \end{vmatrix}$$

For example, $pr_1(a_1)$ is normalized in $pr_1'(a_1) = \dfrac{pr_1(a_1)}{6} = \dfrac{2.373}{6} = 0.395$. Thus, with regard to choosing an MES, a_1 has a priority of 0.395, according to g_1, which is twice as great as a_4 which has a priority of 0.188.

A similar processing can naturally be carried out for all criteria and sub-criteria.

Moreover, the principle of calculating the $pr_i'(a_1)$ is based on ratio averaging operations.

This type of processing does not, however, make it possible to respect preservation of the $di_i(a_j, a_k)$ provided by the decision-maker. Indeed, to identify m unknowns, the $pr_i(a_j)$, AHP uses $\dfrac{m(m-1)}{2}$ independent equations, from $di_i(a_j, a_k)$ which cannot all be respected simultaneously. It can therefore produce what is called rank reversal between the $di_i(a_j, a_k)$ and *ratio* of the $\dfrac{pr_i(a_j)}{pr_i(a_k)}$ calculated, to the extent where $di_i(a_j, a_k) > 1$ and $\dfrac{pr_i(a_j)}{pr_i(a_k)} < 1$ or $di_i(a_j, a_k) < 1$ and $\dfrac{pr_i(a_j)}{pr_i(a_k)} > 1$. In other words, calculating the $pr_i'(a_j)$ can "contradict" the $di_i(a_j, a_k)$ which are at its root [BEL 83; VAR 97]. To limit this risk, the method's author asks, once these calculations have been carried out, for a consistency analysis to be a carried out between the initial $di_i(a_j, a_k)$ and the $pr_i'(a_j)$ found by the calculating procedure. In this, AHP introduces an inconsistency coefficient (*Consistency Index*) written as *CI*. The idea is that this coefficient, positive by definition, should be the weakest possible; a value of $CI \leq 0.1$ is generally considered satisfactory, a value of 0 meaning that all the $di_i(a_j, a_k)$ will have been preserved. In the case where *CI* is too high, the decision-maker is encouraged to reconsider certain $di_i(a_j, a_k)$. The idea is naturally to carry out this verification for all the $pr_i{'}(a_j)$. Consequently, any decision-maker could question the legitimacy of this method, making them review their copy so that the calculations may be rounded!

The AHP method proposes the following formula for *CI*:

$$CI = \frac{\sum_{j=1}^{m} \frac{pr_j{'}(a_j)}{di{'}(a_j, a_j)} - m}{m-1}$$

In fact, it involves checking that the terms of the diagonal of the normalized relative magnitude matrix $di_i{'}(a_j, a_k)$ are close to the value 1.

We note that when all the $\frac{pr_i{'}(a_j)}{di_i{'}(a_j, a_j)}$ are equal to 1, then $CI = \frac{m-m}{m-1} = 0$.

For the sake of clarity, the calculation of *CI* is presented using the example.

To test the coherence of the $pr_1{'}(a_j)$ priorities obtained in regard to the given $di_1(a_j, a_k)$, the decision-maker will be able to apply the formula:

$$CI = \frac{\frac{0.395}{0.41} + \frac{0.066}{0.05} + \frac{0.170}{0.16} + \frac{0.188}{0.17} + \frac{0.094}{0.09} + \frac{0.087}{0.09} - 6}{5} = \frac{6.498 - 6}{5} = 0.099$$

The value $CI = 0.099 \leq 0.1$ would not therefore call the $di_1(a_j, a_k)$ into question.

4.3.3.3. *Acquiring relative magnitudes to calculate the weight*

In AHP, the weights w_i and w_{i_k} $(i,k = 1,...,n)$ reflect respectively the importance of g_i and g_{i_k} with regard to the goal. The method develops these weights in similar fashion to calculating the $pr_i{}'(a_j)$. The decision-maker this time expresses the relative magnitudes $di(g_i, g_p)$ $(i, p = 1,...,n)$ and if necessary, the relative magnitudes $di(g_{i_k}, g_{i_p})$ $(k, p = 1,...,q)$.

> The information system manager will be able to express the relative magnitudes for the sub-criteria of the *Response_to_specifications* criterion:
> – the *Virtualization* sub-criterion is of *moderate* relative magnitude compared to the *Database* sub-criterion: $di(g_{2_1}, g_{2_2}) = 3$ and $di(g_{2_2}, g_{2_1}) = \frac{1}{3}$;
> – the *Database* sub-criterion is of *moderate* relative magnitude than the *Interface* sub-criterion: $di(g_{2_2}, g_{2_3}) = 3$ and $di(g_{2_3}, g_{2_2}) = \frac{1}{3}$;
> – the *Virtualization* sub-criterion is of *stronger* magnitude than the *Interface* sub-criterion: $di(g_{2_1}, g_{2_3}) = 5$ and $di(g_{2_3}, g_{2_1}) = \frac{1}{5}$.
>
> The set of relative magnitudes is assembled in the following matrix:
>
> $$\begin{bmatrix} 1 & 3 & 5 \\ 1/3 & 1 & 3 \\ 1/5 & 1/3 & 1 \end{bmatrix}$$

At this level of knowledge (relative magnitudes of the weights in the form of ratios), it is a matter of calculating the weights of the criteria and sub-criteria. Just as for calculating priorities, AHP proposes providing these weights in a normalized manner. The method proceeds in three stages to do this: normalization, calculation then normalization, in conformity with what is written below for the criteria (the principle being the same for the sub-criteria):

– normalization of the relative magnitudes $di'(g_i,g_p) = \dfrac{di(g_i,g_p)}{\sum_{i=n} di(g_i,g_p)} = 1,$

$\forall p = 1,...,n$, with $di'(g_i,g_p)$ the normalized relative magnitudes;

– calculating the weights $w_i = \sum_{p=1}^{n} di'(g_i,g_p)$;

– normalization of the weights $w_i' = \dfrac{w_i}{n}$ $\forall i = 1,...,n$ $w_i' = \dfrac{w_i}{n}$, with w_i' being the normalized weights.

To obtain the weights of the sub-criteria of the *Response_to_specifications* criterion, the information system manager will begin by normalizing their $di(g_{2_k}, g_{2_p})$, and obtaining the following comparison matrix, where the results of the calculations are rounded to the third decimal:

$$\begin{bmatrix} 0.652 & 0.692 & 0.556 \\ 0.217 & 0.231 & 0.333 \\ 0.130 & 0.077 & 0.111 \end{bmatrix}$$

For example, $di(g_{2_1}, g_{2_2}) = 3$ is normalized in $di'(g_{2_1}, g_{2_2}) = \dfrac{3}{4.33}$ given that $\sum_{k=1}^{n} di(g_{2_k}, g_{2_2}) = 3 + 1 + \dfrac{1}{3} = 4.33$.

The information system manager will therefore obtain:

– $w_{2_1} = 1.900$;

– $w_{2_2} = 0.781$;

– $w_{2_3} = 0.318$.

For example, w_{2_1} is obtained by the sum of the $di'(g_{2_1}, g_{2_k})$ $(k=1,...,3)$: $w_{2_1} = 0.652 + 0.692 + 0.556 = 1.900$.

Finally, the w_{2_k} will be normalized. Therefore:

- $w_{2_1}' = 0.633$;
- $w_{2_2}' = 0.260$;
- $w_{2_3}' = 0.106$.

For example, w_{2_1} is normalized in $w_{2_1}' = \dfrac{w_{2_1}}{3} = \dfrac{1.900}{3} = 0.633$. Thus, with regard to choosing the MES, $w_{2_1}' = 0.633$ is six times greater than $w_{2_3}' = 0.106$.

A similar processing will naturally be carried out for the criteria.

An analysis of the coherence of the $di(g_{i_k}, g_{i_p})$ is then carried out by adapting the formula used for the coherence of the priorities:

$$CI = \frac{\sum_{i=1}^{n} \dfrac{w_i'}{di'(g_i, g_i)} - n}{n-1}$$

To test the coherence of the weights obtained with regard to the relative magnitudes provided, the decision-maker will be able to apply the formula. Thus, concerning the coherence of the weights of the sub-criteria of g_2:

$$CI = \frac{\dfrac{0.633}{0.65} + \dfrac{0.260}{0.23} + \dfrac{0.106}{0.11} - 3}{2} = \frac{3.055 - 3}{2} = 0.028$$

The value $CI = 0.028 \leq 0.1$ will not therefore call the relative magnitudes into question.

4.3.3.4. *Aggregating priorities*

It is now a case of calculating the global priorities of the alternatives. To do this, the $pr_i{'}(a_j)$ and $pr_{i_k}{'}(a_j)$ determined previously are aggregated twice (see section 4.3.3.1):

– the $pr_{i_k}{'}(a_j)$ priorities are first aggregated, by means of the WAM, to calculate $pr_i{'}(a_j)$: $pr_i{'}(a_j) = \sum_{k=1}^{q} w_{i_k} \times pr_{i_k}{'}(a_j)$;

– the $pr_i{'}(a_j)$ priorities are then aggregated by means of the WAM, into a global priority $PR(a_j)$: $PR(a_j) = \sum_{i=1}^{n} w_i \times pr_i(a_j)$.

We note that in this calculation, the $pr_i{'}(a_j)$ are obtained in conformity with the procedure presented previously (see section 4.3.3.2) if the criterion is not broken down, and in conformity with the first stage of the calculation proposed above in the opposite case.

First, the information system manager will be able to obtain the priority according to the *Response_to_specifications* criterion. Therefore, for a_3:

$pr_2{'}(a_3) = 0.633 \times 0.141 + 0.260 \times 0.456 + 0.106 \times 0.310 = 0.241$

Second, the information system manager will be able to calculate the overall priority $PR(a_3)$ attached to a_3 from the $pr_i{'}(a_3)$:

$PR(a_3) = 0.226 \times 0.170 + 0.483 \times 0.241 + 0.185 \times 0.120 + 0.071 \times 0.135$
$+ 0.063 \times 0.293 = 0.197$

Figure 4.8 proposes an overview of the weights and priorities. Since the results concern the single alternative a_3, the notations $pr_i{'}(a_3)$ have been lightened by $pr_i{'}$.

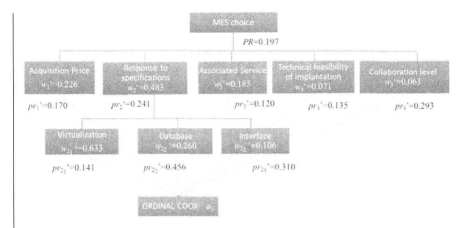

Figure 4.8. *Details of information for calculating PR(a_3) in AHP*

The same calculation should be made for all the alternatives. The information system manager will be able to obtain a total order, as the Hasse diagram in Figure 4.9 shows. The global priorities have been plotted on the diagram. Their choice problem would therefore be solved, less so the guarantee of the results' consistency!

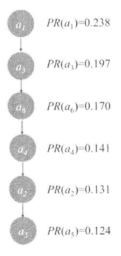

Figure 4.9. *Hasse diagram in the AHP method*

4.3.3.5. *Strengths and limitations*

The decision support provided by AHP thus consists of recommending choosing one alternative. The decision-maker has priorities and weights that allows them to understand the order proposed. However, the information they have was intended for interpretations other than those of superiority (inferiority). This will form both the method's major advantage and its no less major drawback. Indeed, the strength of the AHP method is the total order that can be deduced immediately from the procedure and its weakness will be the significance of this order, in the sense where comparisons between priorities, in the form of difference or *ratio* for example, can only be interpreted by the decision-maker.

In conclusion, the strengths associated with the AHP method are the hierarchy of criteria, the procedure's simplicity for processing relative magnitudes that require only arithmetical calculations, and the quantification of global priorities that allow the establishment of a total order. We might, however, question how well-founded a systematically analytical approach is. It is this approach to which the methods of classical theory have habituated us (see section 2.5). Yet in its spirit, the theory of procedural rationality (from which, to remind you, MCDA is drawn) is intended precisely to address complex problems and so is subject, in essence, to interactions and internal dependencies. This being the case, the main limitation linked to AHP is, as seen above, the mathematical consistency of the procedure for developing weights and priorities, which can lead to a reversal between the relative magnitudes given by the decision-maker and the priorities or weights obtained. This type of situation is illustrated using our example.

Imagine that a new decision-maker, the information system manager's successor, may be asked about their relative magnitudes $di_1(a_j, a_k)$ between the alternatives a_j $(i,k=1,...,6)$.

$$\begin{bmatrix} 1 & 7 & 3 & 2 & 4 & 7 \\ 1/7 & 1 & \frac{1}{4} & 1 & 1/3 & \frac{1}{2} \\ 1/3 & 4 & 1 & 1 & 2 & 5 \\ 1/2 & 3 & 1 & 1 & 3 & 5 \\ 1/4 & 2 & 1/5 & 1/5 & 1 & 1/2 \\ 1/7 & 2 & 1/2 & 1/2 & 2 & 1 \end{bmatrix}$$

This decision-maker could state, in particular, that a_6 is of *very moderately* greater magnitude than a_5: $di_1(a_6, a_5) = 2$ and $di_1(a_5, a_6) = \frac{1}{2}$.

The mechanism for calculating AHP will give the following $pr_1(a_5)$ and $pr_1(a_6)$:

– $pr_1(a_5) = 0.085$;

– $pr_1(a_6) = 0.074$.

Calculating the priorities will therefore lead to the *ratio* $\frac{pr_1(a_5)}{pr_1(a_6)} = \frac{0.085}{0.074} = 1.149$ in favor of the magnitude of a_5 compared to a_6, which is in contradiction with the initial formulation ($di_1(a_5, a_6) = \frac{1}{2}$). There will therefore have been rank reversal between a_5 and a_6.

Authors critical of the method saw in this problem the consequences of not respecting measurement theory in the procedure. Moreover, this would form a fruitful debate on the foundations of AHP and respect for measurement theory which would lead, aside from propositions for developing the method [SCH 94], to new methods in coherence with measurement theory as is the case for the MACBETH method.

Moreover, since the 1970s, Saaty had sensed the benefit of IT assistance in decision support and proposed the *Expert Choice* software for this purpose. Today, a number of softwares make it possible to carry out calculations induced by the method and to construct graphical representations for priorities and weights. For example, the Superdecisions[8] software is relatively easy to handle and offers numerous functions. Moreover, the decision-maker controlling the spreadsheet will be able to systematize the processing by developing their own calculating sheet or using, here again, online services for multi-criteria decision-aiding (Decision Deck in this case).

---

8 http://www.superdecisions.com/.

4.3.4. *The MACBETH method*

4.3.4.1. *General remarks*

Based on measurement theory [KRA 71], the MACBETH method is able to address choice, ranking, sorting and scoring problems (see section 3.4.3). Proposed in the 1990s by Bana e Costa and Vansnick [BAN 97], the method was born in response to the limitations of the AHP method. It consequently takes up the principles of aggregation and hierarchization but restores the notion of utility and uses linear programming to tackle the decision-maker's comparisons. Even though MACBETH shares some tools (such as linear programming) with optimization methods, its philosophy still remains satisfaction (with the alternatives), without ever seeking a hypothetical optimality. Although the authors mention its use in large institutions, the MACBETH method remains modestly applied [FER 21], much less than the AHP method. Indeed, recognized today for its scientific foundations, the choice to use it requires comprehension of and belief in the incoherencies of the AHP method, reserved for "informed" decision-makers. We note convincing results, however, of its application in an industrial environment[9] [LAU 10; SHA 12; KAR 13].

The MACBETH method thus reprises the AHP method. It advocates its principle of "decomplexifying" the reality of the decision problem, without, however, inciting the decision-maker to focus on each criterion (and its sub-criteria) independently of the other criteria. Aggregation is also retained in the same philosophy. But in the case of MACBETH, the differences of attractiveness are replaced with "relative magnitudes" (which, as a reminder, are replaced in their turn with the notion of preference) and the notion of score, reflecting the attractiveness of each alternative, is replaced with that of priority. Although strictly identical to the notion of utility, a desire of the authors to better position the MACBETH method in relation to the principles of MAUT, mentioned previously, would justify this renaming. The use of the term score rather than attractiveness was without doubt chosen for clarity. The order between the alternatives will therefore conform to the global scores obtained.

---

[9] In our work, the MACBETH method has been applied to the problem of describing industrial performance [CLI 07], to managing logistics chains [BER 07], to choosing project portfolios [CLI 06] and to transforming an activity eco-park [LET 21].

Although the calculation principles appear to be the same as in AHP, the aggregation operator's pre-conditions that the values to be aggregated be "significant" and "commensurable" are required [GRA 05a; BOU 06a]. Indeed, it becomes necessary, and this will be transparent to the decision-maker, to aggregate scores that convey the same semantic and which are defined on scales coherent with the aggregation operator chosen (WAM). For the sake of the consistency of the values obtained, the method advocates developing the scores and weights according to interval scales. This development occurs in two stages: first, a development according to a pre-cardinal scale, then an adjustment on a cardinal scale (see section 4.3.4.1). We note that these two scales give meaning to the difference in the scores and are both interval scales. To construct these scales, the method specifies in particular constructing the scores independently of comparing the alternatives. Indeed, when, as in AHP, the priorities of the alternatives are obtained by comparing them to one another, these priorities change when the set of alternatives evolves, i.e. when new alternatives are considered or alternatives initially chosen are excluded. To avoid this pitfall, MACBETH proposes a basis for making comparisons no longer on the problem's alternatives, but on the particular score values introduced beyond the problem. At minimum, two particular significant values should be identified for each criterion: null attractiveness (minimum score) and total attractiveness (maximum score). The score of one alternative will therefore be deduced by linear interpolation between these extreme values. If the linear model does not suit them, the decision-maker will be able to add some intermediate values, the attractiveness of which they will be able to compare. The said differences of attractiveness will be the result of these comparisons. Here, we will call these particular values "reference values". As for the weights, they are still developed globally in a way equivalent to AHP. However, the decision-maker does not express their preferences directly on the weights but on "fictional alternatives" whose global score can be identified with different weights.

Nevertheless, outside the major difference between the calculations, the method proposes a systematic stage of processing the "compatibility" of the differences of attractiveness. This approach will hence present the advantage of rectifying the tendency that some decision-makers might have to seek the ideal of optimizing all the criteria. MACBETH will in this way embody the very spirit of procedural rationality theory (see section 2.6). Through this stage of returning to the differences of attractiveness expressed, it will also be possible, indirectly, to aid the decision-maker to ensure that these reflect their true intention. Similarly, the scores and weights obtained remain subject, in the last stage, to the decision-maker's validation. The decision-maker will be able to judge, at this level too, the conformity of the problem's solution with the vision they may have of it, or at least to note its contradiction with the latter. As it is thought of, the MACBETH method thus proposes that the decision-maker consider a "parallel" problem, removed from reality. Will this augment their reasoning? Moreover, will their vision of this problem always be the same? Will the decision-maker's intention remain the same depending on whether they have a real problem or one with little meaning in reality? Will two decision-makers give the same values and above all the same differences between the values?

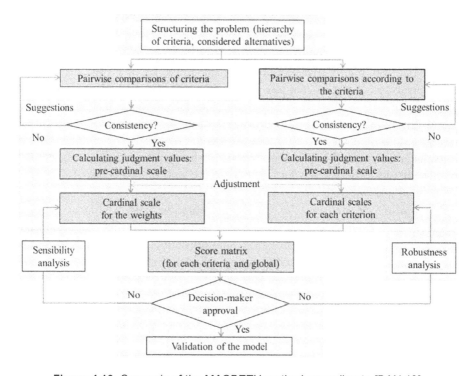

Figure 4.10. *Synopsis of the MACBETH method according to [BAN 12]*

Figure 4.10 presents a synopsis of the method, first positioning the structuring of the decision problem, and then continuing to acquire the differences of attractiveness and potentially to rectify them. The scores and weights are then calculated, then aggregated (using WAM) to calculate the global score. The global scores are then subject to sensitivity and robustness analyses. Still with the same care for harmonization and ease of reading, we will continue to base our analysis, to describe the method, on the notations introduced previously and will make use of screenshots from the M-MACBETH software, which we will mention later.

As for the previous methods, not all stages of the method are detailed. We choose to focus here on the mathematical processes that overcome the limitations of the AHP method as well as on some basic elements to understand the mechanism behind the method. The reader will be able to refer to publications by the method's authors on the different stages of the synopsis [BAN 97; BAN 05]. We will therefore address:

– the acquisition of the differences of attractiveness and calculation of the scores and weights;

– the aggregation of the scores.

MACBETH uses measurement theory to base its calculations, in this instance and as far as we are concerned, on the following precepts:

– the commensurability of the information, i.e. the scores and weights;

– the significance of the aggregation operator, i.e. the WAM.

Commensurability concerns the semantics of the values whereas significance concerns the coherence of aggregating these values with a given operator. Respecting these two conditions makes it necessary to define the information on specific scales [STE 46]. There will therefore be coherence in aggregating the values defined according to ratio scales with operators such as geometric averages, and values defined according to interval scales with operators such as arithmetical averages.

To illustrate these notions, let us consider three alternatives a, b, c and one criterion, g_i. The associated scores will be: $u_i(a)$, $u_i(b)$, $u_i(c)$. These scores will be defined on an interval scale if it is possible to give meaning to their differences, i.e. to compare $(u_i(a) - u_i(b))$, $(u_i(a) - u_i(c))$ and $(u_i(b) - u_i(c))$. This comparison can take the form either of an indifference: $(u_i(a) - u_i(b)) = (u_i(b) - u_i(c))$, for example or of a preference: $(u_i(a) - u_i(b)) > (u_i(b) - u_i(c))$, for example.

4.3.4.2. *Acquiring differences of attractiveness and calculating scores*

Unlike AHP, which is based on comparing alternatives for acquiring what MACBETH calls differences of attractiveness, MACBETH is based on comparing reference values that it has also introduced. These reference values are generally purely fictional values, i.e. they are not taken by the criteria and do not therefore correspond to any of the alternatives envisaged. In this, they appear to call on what can sometimes be called fictional situations.

More precisely, the l reference values associated with g_i are defined on G^i (see section 3.4.2) and written as $r_{i,k}$ ($k=1,...,l$ $l \geq 2$). The decision-maker's attraction to the value $r_{i,k}$ will be described using a score written as $u_i(r_{i,k})$. In this sense, the reference value $r_{i,l}$ will correspond to zero attractiveness, $u_i(r_{i,1}) = 0$, MACBETH calls this level "neutral". Similarly, the reference value $r_{i,l}$ will correspond to total attractiveness, $u_i(r_{i,l}) = 1$, MACBETH calls this level "good". These values will correspond to the bounds of the interval $[0,1]$ of the scores associated with the alternatives.

> Considering the *Acquisition_price* when choosing an MES (see section 3.4.2), the information system manager will be able to imagine new situations to which will correspond respectively the value that would attract them the most (8,000 €) and the one that would not attract them at all (50,000 €), while still remaining acceptable. Since the linear model does not suit them, the information system manager will be able to define new reference values and to choose for themselves four additional values for which they will be able to express a difference of attractiveness. The reference values will therefore be:
>
> $-r_{1,1} = 50{,}000$ €;
>
> $-r_{1,2} = 40{,}000$ €;
>
> $-r_{1,3} = 30{,}000$ €;
>
> $-r_{1,4} = 20{,}000$ €;
>
> $-r_{1,5} = 12{,}000$ €;

$$-r_{1,6} = 8{,}000 \ €.$$

Since the reference values are defined, it is a question for the decision-maker of "judging" the difference in their attractiveness[10]. If the decision-maker judges that there is a difference in attractiveness without nevertheless specifying it, it will be considered *positive*. If the decision-maker judges that there is no difference in attractiveness between the reference values, it will be considered *null*. If the decision-maker can qualify a difference in attractiveness as *positive*, MACBETH proposes, in the same spirit as the AHP method, six additional categories: *very weak, weak, moderate, strong, very strong, extreme*. Finally, the decision-maker will be able to judge the difference in attractiveness by means of these eight categories.

It involves comparing the difference in attractiveness of the reference values $r_{i,k}$ $(k=1,\ldots,l)$ taken two by two. This difference in attractiveness is obtained by comparing the scores associated with these reference values. Thus, the difference in attractiveness for the values $r_{i,k}$ $(k, p = 1,\ldots,l)$ corresponds to $u_i(r_{i,k}) - u_i(r_{i,p})$. Each difference in attractiveness is then qualified, depending on its value, by means of one of the categories defined previously, for example, $u_i(r_{i,k}) - u_i(r_{i,p})$: *null* or $u_i(r_{i,k}) - u_i(r_{i,p})$: *moderate*, etc.

According to the *Acquisition_price*, the information system manager will be able to express differences of attractiveness including, for example:

$-u_1(r_{1,6}) - u_1(r_{1,5})$: *moderate*;

$-u_1(r_{1,6}) - u_1(r_{1,4})$: *strong*;

– etc.

---

10 MACBETH considers only differences of attractiveness $(u(a) - u(b))$ such as $(u(a) \geq u(b))$, the potential differences of attractiveness $(u(b) - u(a))$ providing no additional information for later calculations.

Table 4.8 summarizes the differences of attractiveness between the reference values. The color green indicates the level *good* $(r_{1,6})$, the color blue indicates the level *neutral* $(r_{1,1})$.

Acquisition price	8000	12000	20000	30000	40000	50000
8000	null	moderate	strong	v. strong	extreme	extreme
12000		null	moderate	strong	v. strong	extreme
20000			null	moderate	strong	v. strong
30000				null	weak	moderate
40000					null	v. weak
50000						null

Table 4.8. *Differences of attractiveness for the Acquisition_price in MACBETH*

MACBETH therefore uses the property of interval scales to give a sense to the categories of differences of attractiveness. This property uses the order relationship between of the difference of attractiveness categories that increase from *null* difference of attractiveness up to *extreme* difference of attractiveness. So, for example, a *strong* difference of attractiveness should be greater than a *moderate, weak, very weak* or *null* attractiveness and inferior to a *very strong* or *extreme* difference of attractiveness.

In conformity with the differences of attractiveness in Table 4.8:

$- u_1(r_{1,6}) - u_1(r_{1,5}) < u_1(r_{1,6}) - u_1(r_{1,4})$; the *moderate* difference of attractiveness is lower than the *strong* difference of attractiveness;

$- u_1(r_{1,6}) - u_1(r_{1,4}) < u_1(r_{1,6}) - u_1(r_{1,3})$; the *strong* difference of attractiveness is lower than the *very strong* difference of attractiveness;

$- u_1(r_{1,2}) - u_1(r_{1,1}) < u_1(r_{1,3}) - u_1(r_{1,2})$; the *very weak* difference of attractiveness is lower than the *weak* difference of attractiveness;

– etc.

When the decision-maker has judged all the differences, the method verifies the compatibility of these judgments.

> As an illustration of this notion, take the case of three reference values $r_{i,1}$, $r_{i,2}$, $r_{i,3}$ for which the decision-maker expresses the differences of attractiveness, $u(r_{i,2}) - u(r_{i,1})$: *very weak*, $u(r_{i,3}) - u(r_{i,1})$: *weak* and $u(r_{i,3}) - u(r_{i,2})$: *moderate*. These differences of attractiveness are not compatible since the property of interval scales imposes: $u(r_{i,3}) - u(r_{i,1}) \geq u(r_{i,3}) - u(r_{i,2})$. It is therefore necessary that the decision-maker rectifies at least one of these differences of attractiveness, for example, $u(r_{i,3}) - u(r_{i,1})$: *strong*, which would make it possible to respect the condition $u(r_{i,3}) - u(r_{i,1}) > u(r_{i,3}) - u(r_{i,2})$.

The authors of the method highlight the use of two linear programs, one to verify it and the other to correct the compatibility of the differences of attractiveness expressed.

At this level of knowledge (reference values, difference of attractiveness), it is now a question of calculating, for each criterion, the scores of the associated reference values. This calculation will involve the intermediate values, since the scores of the reference values of the extreme value have already been defined (0 for *neutral* and 1 for *good*). As for AHP, the calculation is normalized and directly involves the criteria when they are not broken down into sub-criteria in the opposite case. MACBETH proceeds to do this in two stages. It first defines the pre-cardinal scales, i.e. each score is bounded by two values, then a cardinal scale by adjusting this score to a single value.

The MACBETH mechanism uses a third, linear program to define the pre-cardinal scales associated with the criteria. It involves minimizing, to obtain the lower bound, and maximizing, to obtain the upper bound, the score associated with each reference value $u_i(r_{i,k})$ ($k = 2,...,l$), under the following constraints:

– the scores of reference values are positive: $u_i(r_{i,k}) \geq 0$ ($k = 1,...,l$);

– the scores are normalized: $u_i(r_{i,1}) = 0$ and $u_i(r_{i,l}) = 1$;

– the differences of attractiveness are ordered: $u_i(r_{i,k}) - u_i(r_{i,p}) < u_i(r_{i,r}) - u_i(r_{i,s})$ or $u_i(r_{i,k}) - u_i(r_{i,p}) > u_i(r_{i,r}) - u_i(r_{i,s})$ or $u_i(r_{i,k}) - u_i(r_{i,p}) = u_i(r_{i,r}) - u_i(r_{i,s})$, where $r_{i,k}$, $r_{i,p}$, $r_{i,r}$, $r_{i,s}$ are reference values.

The information system manager will be able to determine bounds associated with $r_{1,5} = 12,000 $ €, indicated as a percentage in Figure 4.11: $[70.60; 82.34]$. The scale on the right indicates this interval (in red) for $r_{1,5}$, the curve on the left (linear, in red dotted lines) indicates the interval for the values comprised between $r_{1,4} = 20,000$ € and $r_{1,6} = 8,000$ €.

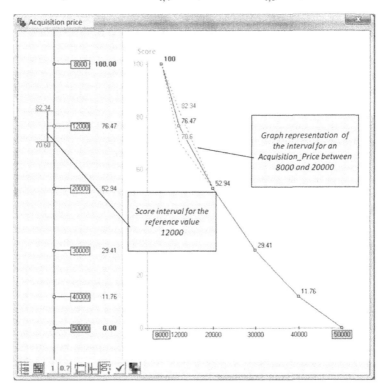

Figure 4.11. *Pre-cardinal scale for the Acquisition_price in MACBETH*

For each reference value included within these two bounds, MACBETH proposes adjusting the score using an approach involving the proportionalities between the differences of attractiveness. The method, however, offers the decision-maker the possibility of proposing their own adjustment, on condition that the score remains within the interval associated with it. In all cases, this cardinal scale naturally verifies the previous constraints and so remains an interval scale. Once this adjustment is made, it is possible to deduce the score of the alternatives through linear interpolation between the two closest reference values each time. So, by considering r_{ik}, r_{ik+1} which are the reference values closest to value c_{ij} the score for alternative a_j will be obtained in the following way:

$$u_1(a_j) = u_1(r_{ik+1}) + (u_1(r_{ik+1}) - u_1(r_{ik})) \times \frac{c_{ij} - r_{ik}}{r_{ik+1} - r_{ik}}$$

The information system manager could conserve the values adjusted using MACBETH for the *Acquisition_price* according to the cardinal scale in Figure 4.12 or proceed with their own adjustment. They could then deduce the scores $u_1(a_j)$ of the alternatives.

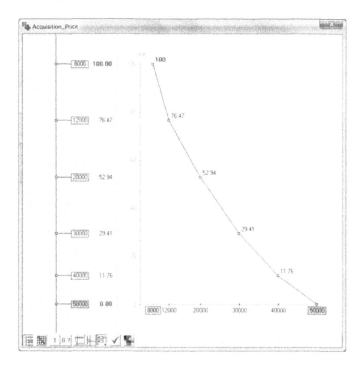

Figure 4.12. *Cardinal scale for the Acquisition_price in MACBETH*

Thus, the score of a_2 for which the *Acquisition_price* is 34,000 € and for which the closest reference values are 30,000 € and 40,000 € will be:

$$u_1(a_2) = 0.1176 + (0.2941 - 0.1176) \times \frac{34{,}000 - 30{,}000}{40{,}000 - 30{,}000} = 0.1882.$$

4.3.4.3. *Acquisition of differences of attractiveness, calculating weights*

To calculate the weights, the method is based on expressing differences of attractiveness, just as for calculating the scores. Unlike AHP, the authors defend the difficulty of comparing the weights to one another, since these are simple artifacts without their own reality. The idea here is to compare situations, and more precisely the global scores associated with them, since these scores link the weights to one another. Another difference with AHP lies in the fact that all the criteria and sub-criteria are considered simultaneously, thus instituting a single and overall aggregation stage, rather than a cascading aggregation (see section 4.3.3.1).

Moreover, for the requirements of linear programming and simplifying processes, the method introduces particular situations, called "fictitious alternatives". In the first place, two fictitious alternatives corresponding to global scores 0 and 1 should be defined:

– the fictitious alternative "$\alpha_{all.\inf.}$" is associated with the global score 0, $U(\alpha_{all\_\inf.}) = 0$, i.e. the score associated with each criterion or sub-criterion is zero, $u_i(\alpha_{all\_\inf.}) = u_{i_k}(\alpha_{all\_\inf.}) = 0$ [11];

– the fictitious alternative "$\alpha_{all.\sup.}$" is associated with the global score 1, $U(\alpha_{all\_\sup.}) = 1$, i.e. the score associated with each criterion or sub-criterion is 1 $u_i(\alpha_{all\_\sup.}) = u_{i_k}(\alpha_{all\_\sup.}) = 1$.

In the second place, a fictitious alternative α_i $(i=1,\ldots,n)$ $(\alpha_{i_k}\ (k=1,\ldots,q))$ is associated with each g_i (g_{i_k}), so that the score

---

11 To avoid any confusion with the alternatives considered in the decision, the fictitious alternatives are written as α.

according to g_i (g_{i_k}) corresponds to the level *good*, $u_i(\alpha_i)=1$ $\left(u_{i_k}(\alpha_{i_k})=1\right)$ when the scores associated with the other criteria and sub-criteria correspond to the level *neutral*, $u_p(\alpha_i)=0$ $(p \neq i, p=1,...,n)$ $\left(u_{i_p}(\alpha_{i_k})=0 \ (p \neq k, p=1,...,q)\right)$.

When choosing their MES, the information system manager will be able to consider the following fictitious alternatives:

– α_1 for which the score according to g_1 (*Acquisition_price*) is 1, $u_1(\alpha_1)=1$ and the scores according to all the other criteria (*Associated_service*, *Technical_feasibility_of_implementation* and *Collaboration_level*) and sub-criteria (*Virtualization*, *Database* and *Interface*) are 0, $u_3(\alpha_1)=0 .... = u_{2_1}(\alpha_1)=...=0$;

– α_{2_1} for which the score according to g_{2_1} (*Virtualization*) is 1, $u_{2_1}(\alpha_{2_1})=1$ and the scores according to all the other criteria (*Acquisition_price*, *Associated_service*, *Technical_feasibility_of_ implementation* and *Collaboration_level*) and sub-criteria (*Database* and *Interface*) are 0, $u_{2_1}(\alpha_1)=...=u_{2_1}(\alpha_{2_2})=...=0$;

– $\alpha_{all.inf.}$ for which the scores according to all the criteria and sub-criteria are 0, $u_1(\alpha_{all.inf.})=u_3(\alpha_{all.inf.})=u_4(\alpha_{all.inf.})=u_5(\alpha_{all.inf.})=u_{2_1}(\alpha_{all.inf.})$
$=u_{2_2}(\alpha_{all.inf.})=u_{2_3}(\alpha_{all.inf.})=0$

– etc.

It can therefore be noted that the expression of the global score associated with α_i (α_{i_k}) can be identified with the weight w_i (w_{i_k}). For the sake of simplicity, this identification will only be illustrated in the example.

The information system manager will be able to identify the global score of α_1 associated with the *Acquisition_price* criterion:

$$U(\alpha_1) = w_1 \times u_1(\alpha_1) + w_3 \times u_3(\alpha_1) + w_4 \times u_1(\alpha_1) + w_5 \times u_3(\alpha_1)$$
$$+ w_{2_1} \times u_{2_1}(\alpha_1) + w_{2_2} \times u_{2_2}(\alpha_1) + w_{2_2} \times u_{2_2}(\alpha_1) = w_1$$

where the first term is equal to w_1 $(u_1(\alpha_1) = 1)$, and all the others are zero. They will be able to do the same for the three other criteria and the three sub-criteria.

Once the fictitious alternatives have been defined, it will then fall to the decision-maker to judge the difference in their attractiveness between the global scores associated with them by reprising the same mechanism as for the scores for the reference values.

To determine the weight, the information system manager will be able to express the differences of attractiveness between the fictitious alternatives considered. They will express a *very weak* difference of attractiveness, between α_1 (fictitious alternative associated with the *Acquisition_price*) and α_4 (fictitious alternative associated with the *Technical_feasibility_of_ implementation*), which means that there are very few differences between the weights of these criteria. They will be able to do the same for all the fictitious alternatives:

– $U(\alpha_1) - U(\alpha_4)$: *very weak*;

– $U(\alpha_1) - U(\alpha_{2_1})$: *moderate*;

– $U(\alpha_4) - U(\alpha_{2_1})$: *positive*;

– $U(\alpha_{2_2}) - U(\alpha_{2_1})$: *null*;

– etc.

Table 4.9 summarizes the differences of attractiveness between the fictitious alternatives. These are designated by the initials of the associated criterion. The last line and column in blue corresponding to $\alpha_{all.inf}$.

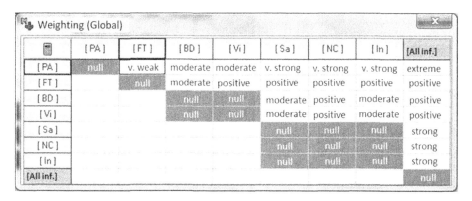

Table 4.9. *Differences of attractiveness between the fictitious alternatives in MACBETH*

As for the scores, MACBETH uses the property of interval scales to give a meaning to these differences of attractiveness.

In conformity with the differences of attractiveness in Table 4.9, the transcription of the categories of degrees attractiveness will be written as follows:

- $U(\alpha_1) - U(\alpha_{2_1}) < U(\alpha_1) - U(\alpha_5)$; the *moderate* difference of attractiveness between α_1 and α_{2_1} is lower than the *very strong* difference of attractiveness between α_1 and α_5;

- $U(\alpha_5) - U(\alpha_{2_1}) > 0$; the difference of attractiveness is *positive*;

- $U(\alpha_{2_2}) - U(\alpha_{2_1}) = 0$; the difference of attractiveness is *null*;

- etc.

As for the differences of attractiveness concerning the reference values, the method verifies and proposes, if applicable, rectifying the differences of attractiveness affecting the fictitious alternatives thanks to the two linear programs mentioned previously.

At this level of knowledge (fictitious alternatives, difference of attractiveness between fictitious alternatives), it is now a question of calculating the global score of the fictitious alternatives and deducing the weight from it. The calculation is normalized and directly concerns the criteria and sub-criteria. As for the scores, MACBETH proceeds in two stages. It first defines a pre-cardinal scale, i.e. the score is bounded by two values, then a cardinal scale, i.e. the score is adjusted to a single value.

The MACBETH mechanism uses a new linear program to define the pre-cardinal scale associated with the weights. It involves respectively minimizing the global score, to obtain the lower bound, and maximizing the global score $U(\alpha_i)$ $(i=1,...,n)$ to obtain the upper bound, under the following constraints:

– normalizing the weights: $\sum_{i=1}^{n} w_i = 1$;

– defining the fictitious alternative $\alpha_{all.inf.}$: $U(\alpha_{all.inf.}) = 0$;

– the differences of attractiveness: $U(\alpha_i) - U(\alpha_k) < U(\alpha_r) - U(\alpha_s)$ or $U(\alpha_i) - U(\alpha_k) = U(\alpha_r) - U(\alpha_s)$, where $\alpha_i, \alpha_k, \alpha_r, \alpha_s$ $(i,k,r,s = 1,...,n)$ are the fictitious alternatives considered.

The information system manager will be able to determine the bounds of the global score associated with α_{2_1} (fictitious alternative associated with *Virtualization*) with which w_{2_1} is identified, indicated as a percentage in Figure 4.13: $[13.48; 17.22]$. They can also observe that for *null* differences of attractiveness between fictitious alternatives, the associated weights are identical.

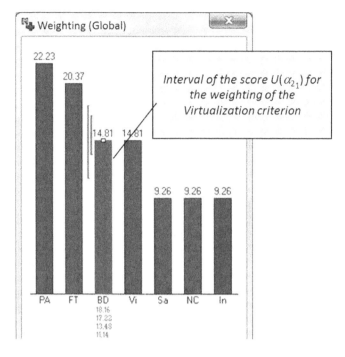

Figure 4.13. *Pre-cardinal scale for weights in MACBETH*

In the same way as for the scores, MACBETH adjusts the weight using an approach involving the proportionalities between the differences of attractiveness. The method, however, offers the decision-maker the chance to propose their own adjustment, depending on how they feel, by nevertheless imposing the sole condition that the score remains within the interval associated with it. As for the scores, this cardinal scale naturally verifies previous constraints and so remains an interval scale.

The information system manager will be able to conserve the values adjusted by MACBETH for the weight of the criteria according to the cardinal scale in Figure 4.14, or proceed to their own adjustment. The most important weight is that of the *Acquisition_price* criterion $w_1 = 0.2223$, the least important weights are those of the *Collaboration_level* and *Associated_service* criteria and of the sub-criterion *Interface* $w_{2_3} = w_4 = w_5 = 0.0926$.

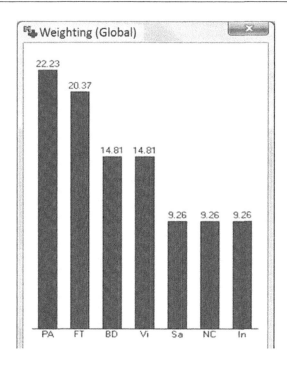

Figure 4.14. *The weights of the criteria and sub-criteria in MACBETH*

4.3.4.4. *Aggregating the scores*

This involves calculating the global scores of the alternatives. To do this, MACBETH simultaneously aggregates the scores of the criteria and sub-criteria by means of the WAM, unlike the AHP method which proceeds to aggregation in multiple stages (with priority of the criteria from the priorities of the sub-criteria then global priority). We note that since the values u_i and u_{i_k}, on the one hand, and the weights w_i and w_{i_k}, on the other hand, are defined according to the interval scales, aggregation by WAM will also form an interval scale where the difference between the global scores associated with the alternatives will also have meaning. For the sake of simplicity, this calculation will only be illustrated using the example.

The information system manager will be able to obtain the global scores of the alternatives directly. Thus, considering a_3: $ORDINAL\_COOX$, they will use WAM to aggregate the associated scores, given as a percentage in the red box in Table 4.10, knowing the weights of the criteria, also given as a percentage in the last line of the same table. The global score will be:

$$U(a_3) = 0.2223 \times 0.7814 + 0.2937 \times 0.4286 + 0.0926 \times 0.6667 + 0.0926$$
$$\times 0.800 + 0.1481 \times 0.6667 + 0.1481 \times 0.8333 + 0.0926 \times 0.79059 = 0.684$$

Score table

Options	Global	PA	Vi	BD	In	FT	SA	NC
XL	46.73	94.67	50.00	16.67	52.94	0.00	77.78	40.00
Aq	51.30	26.67	83.33	50.00	52.94	71.43	0.00	66.67
OC	68.43	78.14	66.67	83.33	70.59	42.86	66.67	80.00
Wo	63.97	87.96	33.33	66.67	58.82	42.86	100.00	66.67
Qu	51.95	73.96	33.33	16.67	35.29	71.43	44.44	66.67
CC	37.70	69.27	16.67	33.33	76.47	0.00	44.44	40.00
[All sup.]	100.00	100.00	100.00	100.00	100.00	100.00	100.00	100.00
[All inf.]	0.00	0.00	0.00	0.00	0.00	0.00	0.00	0.00
Weights		0.2223	0.1481	0.1481	0.0926	0.2037	0.0926	0.0926

Table 4.10. *The scores in MACBETH*

A total order is obtained, as the Hasse diagram shows in Figure 4.15, on which the global scores have been plotted. The problem of choosing will thus be solved.

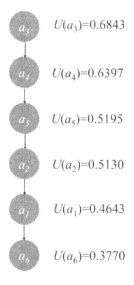

Figure 4.15. *Hasse diagram in the MACBETH method*

The results obtained can then be reconsidered by the decision-maker through a sensitivity study (which concerns the evolution of the score and thus of the rank of the alternatives when the weights are otherwise adjusted) and robustness study (which concerns the evolution of the score and thus of the rank of the alternatives when the scores are otherwise adjusted).

4.3.4.5. *Strengths and limitations*

As for AHP, the decision-aiding provided by MACBETH consists of recommending the choice of one alternative. The decision-maker has scores and weights that enable them to understand the order proposed and to give a meaning to their differences; the foundation of this order and these differences conforming to measurement theory. This is really the method's major asset, these interval scales that, by guaranteeing the consistency of the scores and weights, make it possible to address scoring problems, aside from problems of sorting and choosing (see section 3.4.3). This knowledge makes MACBETH a method that integrates questioning of the results obtained, given the possible adjustments of the weights (the sensitivity study) and scores (the robustness study). Finally, we indicate that this definition of scores and weights according to the interval scales also makes it possible to consider any type of additive aggregation operator, beyond WAM. This will be the case of the Choquet integral, for example, which makes it possible to refine the definition of the interactions between the different criteria [CLI 04; GRA 05b; MAY 10]. Choosing such an operator makes it possible to consider the importance of each criterion considered in isolation, and also the importance of the criteria considered jointly (by twos, by threes, etc.). In addition, in its two-additive version, the Choquet integral can be written as a weighted average corrected by interactions with the criteria taken two by two, which can limit or on the contrary amplify the compensation effects between criteria. In other words, beyond weighting the criteria, this type of operator makes it possible to better translate a decision-maker's intentions, depending on the level of their requirements (the interactions will reduce or increase the compensation) or its counterpart for the compromise (there will be no interactions and the compensation will be that of the weighted average).

This being the case, the main limitation of MACBETH lies in its complexity and its principle, which could seem distanced from intuition. The notion of an interval scale remains slight and requires respecting many constraints: defining the bounds of the scale, guaranteeing the meaning given to difference and ensuring the commensurability of scales. The desire to define these scales independently of the alternatives considered by recourse to the notion of reference values requires additional efforts at understanding. Finally, choosing specific vocabulary for the method (pre-cardinal scale, fictitious alternatives, differences of attractiveness) and recourse to linear programming increase this difficulty of comprehension for the user, who must "learn" the method without always understanding what it provides in the form of information or obtains as a result. This limitation is, however, lessened thanks to the interactivity allowed by the M-MACBETH[12] application software, which supports all stages of the method. In this way, the decision-maker who is practiced in the method will become its fervent "supporter", whereas those who merely discover it will retain only the difficulties of implementing it.

4.4. Conclusion

At the end of this chapter, the reader will have discovered how to go further in their quest for a total order of the alternatives from which they wish to choose. To do this, they should retain a method, and most often nuance their preferences and weight the criteria. These "settings" require proper understanding of the mechanism that, as we recognize, requires some learning, doubtless with the exception of the Borda count method.

This choice of method already reflects the decision-maker's intention. Do they want a simple method, the result of which can easily be explained to all stakeholders in the decision? Do they instead want a more subtle method, one that considers the importance of the criteria and the satisfaction associated with each alternative? Or do they even wish to "decomplexify" their problem to make it easier to tackle for themselves and for stakeholders in the decision? Or finally, do they wish to respect measurement theory and to pay the price for this by making the information processing carried out less accessible? The decision-maker may, of course, ask all these questions. But beyond the method, they should not forget that they are using a decision process in which all these methods are recognized.

---

12 M-MACBETH allows MACBETH to be used free of charge in the "academic" version and can address decision problems when only a limited number of criteria and alternatives is involved. A paid-for version does not have this limitation. Information can be accessed at: htttp://macbeth.com/.

Let us venture an aside. After hearing us present an article using MACBETH at a conference, Jacquet-Lagrèze, one of the authors of the UTA method, came to introduce himself to us. We still did not know him and at first thought he had come to discuss the benefit that using UTA might bring to addressing the problem. He was not. He congratulated us on having worked methodically, emphasizing the need to follow the stages of the decision process, relativizing the choice of the method and encouraging us to pursue this path. Everything was said, posing the problem, using a method with known strengths and limitations, taking the results with a pinch of salt and finally observing at what point the application of one method allows us to better understand the problem. We regret that we were not able to present this UTA method in this chapter. Yet nevertheless, how much strength there is in this "disaggregation" method, which plays the role of considering preferences (between alternatives) holistically to deduce the utilities from them (according to the criteria) depending on the interval scales. If the reader wishes to exert their right to know more about MCDA, we encourage them to discover this method.

It is in this spirit that rather than recommending a method to the reader, even though we use some of these in our own work, we recommend that they discover other methods and in particular PROMETHEE, TOPSIS and Robust Ordinal Regression methods, since each of these methods has specific features adapted to particular decision problems. We also prompt them to focus on group decisions. Since there are sometimes multiple decision-makers and since they are often joined by stakeholders who contribute to the process, some methods such as MACBETH or ELECTRE III encourage a collective construction of the decision process more than others. But beyond the method, what counts in MCDA is doubtlessly just as much the learning of the decision problem that allows a method to be applied, as the order obtained after its application, an order that might still be disputed, since it is constructed from the decision-maker's own preferences and priorities.

Conclusion

So, Ready to Decide?

Although Renard asked about the *quality* of his decision, at the risk of losing oneself in this question, this book sought to investigate the act of deciding, how a decision-maker manages to choose, to arbitrate at the risk of no longer knowing how to proceed, since the decision process presents so much richness and subtlety. To avoid getting lost in the labyrinth of concepts, principles, schools and other theories to only end up questioning ourselves, the intention here was to put forward some tangible elements, making it possible to nourish the decision-maker's reflection, a reflection that lies within each of us.

Within this idea, we first invoked history, offering the perspective of the practice of the great civilizations on this question of decision and doing so from their shared aspects (a hierarchized, centralized and formalized vision), as well as in their specifics (ideal, compromise, improvement, etc.). What then appeared as determinant were the beliefs, values, habits, fears, knowledge and other aspects that shape humankind and color their decisions beyond their own intentions. However, it remains that when the decision-maker works on the basis of reason and does not know how to access or leave aside their intuition, they need a method to reach this decision. No doubt it is a bias of the authors to situate themselves in the Western context and the vision of Descartes, to depict here a decision imprinted with logic and rationality. But would we have been able, bathed in our beliefs and values, to explore other ways, such as the *path* belonging to Taoism or the *search for equilibrium* of Buddhism? It is therefore the context of a reflected decision, guided by methods to which these reflections belong.

Please, draw me a satisfactory solution, the apprentice decision-maker therefore asks. And Descartes' successors, probabilists, mathematicians, information specialists and others have pushed the requirements of methods to the extreme for two centuries, to answer: *to decide is to optimize!* But other successors, including economists, possibilists and psychologists, defend a method that hinges more on *how to decide*, to answer: *to decide is to satisfy!* And the apprentice decision-maker finds themselves with two pictures without knowing which to choose. It was the aim of Chapter 2 to characterize both major decision theories and to underline their complementary, rather than competitive character. So long as the decision problem is well framed, and the decision-maker has complete knowledge of it, classical theory will simply say *choose the best adapted optimization method*. In the other cases, procedural rationality theory will say *reflect using the method* and more explicitly, *inform yourself, design solutions, evaluate them and choose one that satisfies you*. The two pictures therefore share a format, monitoring a process, but the role given to the decision-maker and their free will will differ, reduced in the case of optimization and much broader in the case of procedural rationality. Nevertheless, in both cases, there will be a need for real expertise from the decision-maker, whether this lies in choosing the formal method or in respecting the stages of a procedure. It was the picture dealing with the search for a satisfactory solution that was addressed in Chapters 2–4, particularly in view of the often poorly put together decision problems encountered in industrial practice. But would this first positioning suffice for our apprentice decision-maker?

Please, draw me a really satisfactory solution, insists the apprentice decision-maker. It was not Saint-Exupéry who wished it[1], but Chapter 3 sought to sketch this solution in the case of procedural rationality. Situated within the framework of a decision that results from choosing a *satisfactory* alternative from a set of alternatives and relying generally on several criteria, the idea was to establish an order between these alternatives with a simple idea: the lower the rank in this order, the greater the satisfaction. This notion of order calls on a similar notion, that of preference. But alas, this order is often partial, mingling strict preference – which the decision-maker seeks – and indifference or incomparability – which the decision-maker fears. Finally, armed only with the notion of Pareto dominance, the decision-maker

---

1 In *Le Petit Prince*, the famous novella of de Saint Exupéry, the little prince begins the meeting with the pilot lost in the desert (who is also the narrator) by asking him "draw me a sheep".

will still be able to eliminate the alternatives to which others will be strictly preferred. Slight progress for the apprentice decision-maker who has here an approach to the decision. Yet they know that the final work, ordering the alternatives will require investigating their preferences. The door will thus have been opened to methods apt to establish this relationship by inviting the decision-maker to put their stamp on it, by choosing a method, and the information the method will require from them.

And when will the drawing of this solution be achieved? hurries the apprentice decision-maker, impatient for a result. Charcoal, watercolor, oil, Paintbrush, Photoshop: the choice is broad for finalizing the picture, ranging from the pointillism of the school of outranking to the numerical precision of the school of aggregation. Knowing the methods, being aware of the information to be provided and the processing being carried out, knowing the strengths and limitations of this or that method, taking a step back to look at the results, was the subject of Chapter 4. The apprentice decision-maker is already aware that the picture depends on them. They now know that the Condorcet method is inapplicable when 20 or so projects are on the table, that Borda's method is simple with a very fast result but that it presents an incoherence in their aggregation model of the ranks, and they also know many things about ELECTRE III, AHP or MACBETH. In addition, they know how to describe their problem better, understand the "why" of the information to be provided to reach a total order, explain the results obtained and make it a tool for collective discussion rather than imposing a decision made hierarchically. And this is how the satisfactory solution will be sketched naturally, given the particular illumination that the method will have given them.

And the apprentice decision-maker recalled that he was a manager in industry. Ah yes, the decision is exercised in a domain, on an object, in a given context, which will not fail to give them its specifics. What are the types of problems encountered? Should decisions be made in an emergency or in a calm state? Are there other stakeholders, or indeed a community of decision-makers? What are the challenges and risks linked to the decision? What about uncertainty about the objectives and information relating to the problem? etc. We now leave our apprentice to face alone the decisions they will make in the environment, industrial or not. But like a *Compagnon du*

Devoir[2] (apprentice artisan) on their obligatory *Tour de France* at the end of their apprenticeship, they have the knowledge to make their decision in conformity with the canons of the discipline, resorting, if they feel the need, to the explanations presented in this book. Perhaps they could even contribute to reflection on the subject, by bearing witness to their practice in the domain of choosing production management methods, instrumenting the Deming wheel, quantifying dashboards, introducing artificial intelligence in predictive maintenance, etc. and thus forming a gallery of decision dashboards for the industrial environment, much as Monet continued to paint the same cathedral at Rouen throughout different hours and seasons[3]. As for the researcher, strengthened by this new knowledge, they will be able to revisit their design of performance in business, by linking it to the notion of satisfaction, and in particular by adapting aggregation methods to the mechanism for expressing this performance. This work remains to be done and will form a natural sequel to this book.

---

2 The *Compagnons du Devoir* is an organization of artisans that originated in the Middle Ages. During their compulsory *Tour de France*, they learn their future profession from their masters.

3 https://www.rouentourisme.com/monet-et-ses-cathedrales/.

References

[ALS 74] ALSTER B., *The Instructions of Shurrupak: A Sumerian Proverb Collection*, Akademisk Forlag, Copenhagen, 1974.

[ARR 62] ARROW K., *Social Choice and Individual Values*, John Wiley & Sons, New York, 1962.

[ARR 98] ARROW K., *Choix collectifs et préférences individuelles*, Diderot Éditions, Paris, 1998.

[AXE 62] AXELOS K., *Héraclite et la philosophie : la première saisie de l'être en devenir de la totalité*, Éditions de Minuit, Paris, 1962.

[BAK 74] BAKER K.R., *Introduction to Sequencing and Scheduling*, John Wiley & Sons, New York, 1974.

[BAN 90] BANARSIDASS M., *Philosophy in the Samadhiraja Sutra de Konstanty*, Regamey, Delhi, 1990.

[BAN 97] BANA E COSTA C., VANSNICK J.C., "Applications of the MACBETH approach in the framework of an additive aggregation model", *Journal of Multi-Criteria Decision Analysis*, vol. 6, no. 2, pp. 107–114, 1997.

[BAN 05] BANA E COSTA C., DE CORTE J.M., VANSNICK J.C., "On the mathematical foundations of MACBETH", in FIGUEIRA J., GRECO S., EHRGOTT M. (eds), *Multiple Criteria Decision Analysis: State of the Art Surveys*, Springer, Berlin, 2005.

[BAN 12] BANA E COSTA C., DE CORTE J.M., VANSNICK J.C., "MACBETH", *International Journal of Information Technology & Decision Making*, vol. 11, no. 2, pp. 359–387, 2012.

[BEL 57] BELLMAN R.E., *Dynamic Programming*, Princeton University Press, New Jersey, 1957.

[BEL 83] BELTON V., GEAR T., "On a short-coming of Saaty's method of analytic hierarchies", *Omega*, vol. 11, no. 3, pp. 228–230, 1983.

[BEL 02] BELTON V., STEWART T., *Multiple Criteria Decision Analysis: An Integrated Approach*, Springer Science & Business Media, Berlin, 2002.

[BEN 89] BENTHAM J., *Introduction aux principes de morale et de législation*, Centre Bentham, Paris, 1789.

[BER 34] BERNOULLI D., *Recueil des pièces qui ont remporté le prix de l'Académie des Sciences*, Lambert, Paris, 1734.

[BER 07] BERRAH L., CLIVILLÉ V., "Towards an aggregation performance measurement system model in a supply chain context", *Computers in Industry*, vol. 58, no. 7, pp. 709–719, 2007.

[BER 13] BERTHOZ A., *La Décision*, Odile Jacob, Paris, 2013.

[BER 15] BERRAH L., CLIVILLÉ V., "Decision making in a coordinated control structure", *International Journal of Multicriteria Decision Making*, vol. 5, no. 1, pp. 39–58, 2015.

[BHA 96] BHAKTIVEDANTA SWAMI PRABHUPADA A.C., *La Sri Isopanisad*, The Bhaktivedanta Book Trust, Mumbai, 1996.

[BOR 14] BORDREUIL P., BRIQUEL-CHATONNET F., MICHEL C., *Les débuts de l'Histoire – Civilisations et cultures du Proche-Orient ancien*, Éditions Kheops, Paris, 2014.

[BOR 21] BOREL E., "La théorie du jeu et les équations intégrales à noyau symétrique", *Comptes rendus hebdomadaires des séances de l'Académie des sciences*, vol. 173, pp. 1304–1308, 1921.

[BOT 97] BOTÉRO J., *Mésopotamie : l'écriture, la raison et les dieux*, Folio, Paris, 1997.

[BOU 06a] BOUYSSOU D., DUBOIS D., PIRLOT M. et al., *Concepts et méthodes pour l'aide à la décision, volume 1 : Outils de modélisation*, Hermès-Lavoisier, Paris, 2006.

[BOU 06b] BOUYSSOU D., DUBOIS D., PIRLOT M. et al., *Concepts et méthodes pour l'aide à la décision, volume 2 : Risque et incertain*, Hermès-Lavoisier, Paris, 2006.

[BOU 06c] BOUYSSOU D., DUBOIS D., PIRLOT M. et al., *Concepts et méthodes pour l'aide à la décision, volume 3 : Analyse multicritère*, Hermès-Lavoisier, Paris, 2006.

[BOU 14] BOUYAUX P., "Économétrie et théorie des jeux", available at: https://www.techniques-ingenieur.fr/base-documentaire/sciences-fondamentales-th8/applications-des-mathematiques-42102210/econometrie-et-theorie-des-jeux-af1500/, 2014.

[BRA 81] BRAUDEL F., *Civilization and Capitalism 15th–18th Century: Three Volumes*, Collins, Glasgow, 1981.

[BRA 02] BRANS J.-P., MARESCHAL B., *PROMETHEE-GAIA. Une Méthodologie d'aide à la décision en présence de critères multiples*, Ellipses, Paris, 2002.

[BRA 05] BRANS J.-P., MARESCHAL B., "Promethee methods", in FIGUEIRA J., GRECO S., EHRGOTT M. (eds), *Multiple Criteria Decision Analysis. State of the Art Surveys*, Springer, Berlin, 2005.

[BRA 10] BRAGGE J., KORHONEN P., WALLENIUS H. et al., "Bibliometric analysis of multiple criteria decision making/multiattribute utility theory", in EHRGOTT M., NAUJOKS T.S., WALLENIUS J. (eds), *IXX International MCDM Conference Proceedings*, Springer, Berlin, 2010.

[BRA 13] BRAUDEL F., *Grammaire des civilisations*, Flammarion, Paris, 2013.

[BRI 05] BRISSON L., *Platon, Criton*, Flammarion, Paris, 2005.

[BUR 99] BURKE L.A., MILLER M.K., "Taking the mystery out of intuitive decision making", *The Academy of Management Executive*, vol. 13, no. 4, pp. 91–99, 1999.

[CAR 61] CARROLL C.W., "The created response surface technique for optimizing nonlinear, restrained systems", *Operations Research*, no. 9, pp. 169–184, 1961.

[CHA 55] CHARNES A., COOPER W.W., FERGUSON R.O., "Optimal estimation of executive compensation by linear programming", *Management Science*, vol. 1, no. 2, pp. 138–151, 1955.

[CHA 93] CHANCHEVIER M., "L'ingénierie simultanée, un nouveau mode de management de projets", *Conférence AFITEP*, Paris, November 1993.

[CHA 06] CHAUVEL A.M., *Méthodes et outils pour résoudre un problème – 55 outils pour améliorer les performances de votre entreprise*, Dunod, Paris, 2006.

[CHE 89] CHENG A., "Un Yin, un Yang, telle est la Voie : les origines cosmologiques du parallélisme dans la pensée chinoise", *Extrême-Orient, Extrême-Occident*, no. 11, pp. 35–43, 1989.

[CHU 57] CHURCHMAN C.W., ACKOFF R.L., ARNOFF E.L., *Introduction to Operations Research*, John Wiley & Sons, New York, 1957.

[CHU 89] CHU C., PORTMANN M.-C., Minimisation de la somme des retards pour les problèmes d'ordonnancement à une machine, Research report, INRIA, 1989.

[CLI 04] CLIVILLÉ V., Approche systémique et méthode multicritère pour la définition d'un système d'indicateurs de performance, PhD thesis, Université de Savoie Mont Blanc, Chambéry, 2004.

[CLI 06] CLIVILLÉ V., BERRAH L., "Une approche multicritère pour l'aide à la sélection de portefeuilles de projets", *Conférence Francophone de MOdélisation et SIMulation (MOSIM'06)*, Rabat, 2006.

[CLI 07] CLIVILLÉ V., BERRAH L., MAURIS G., "Quantitative expression and aggregation of performance measurements based on the MACBETH multi-criteria method", *International Journal of Production Economics*, vol. 105, no. 1, pp. 171–189, 2007.

[CLI 13] CLIVILLÉ V., BERRAH L., MAURIS G., "Deploying the ELECTRE III and MACBETH multicriteria ranking methods for SMEs tactical performance improvements", *Journal of Modelling in Management*, vol. 8, no. 3, pp. 348–370, 2013.

[CNR 21] CNRTL, "Centre National de Ressources Textuelles et Lexicales", available at: https://www.cnrtl.fr/definition/, 2021.

[CON 85] CONDORCET (CARITAT MARQUIS DE CONDORCET) J., *Essai sur l'application de l'analyse la probabilité des décisions rendues à la pluralité des voix*, Imprimerie Royale, Paris, 1785.

[COU 38] COURNOT A., *Recherches sur les principes mathématiques de la théorie des richesses*, Hachette, Paris, 1938.

[CRO 63] CROUZET M., *Histoire générale des civilisations*, 7 volumes, Presses universitaires de France, Paris, 1963.

[DAM 94] DAMASIO A., *L'erreur de Descartes : la raison des émotions*, Odile Jacob, Paris, 1994.

[DAN 48] DANTZIG G.B., Programming in a linear structure, USAF Project Rand, Document, USAF, 1948.

[DAN 90] DANTZIG G.B., "Origins of the simplex method", in NASH S.G. (ed.), *A History of Scientific Computing*, Pearson, Reading, 1990.

[DEB 81] DE BORDA J.-C., *Mémoire sur les élections au scrutin*, Histoire de l'Académie royale des sciences, Paris, 1781.

[DEB 83] DEBREU G., HILDENBRAND W., *Topological Methods in Cardinal Utility Theory*, Cambridge University Press, Cambridge, 1983.

[DEC 98] DECREUSE C., FESCHOTTE D., "Ingénierie simultanée", *Techniques de l'ingénieur*, 1998.

[DEM 82] DEMING E., *Quality, Productivity and Competitive Position*, MIT Press, Cambridge, 1982.

[DEM 05] DEMAILLY A., PINGAUD F., "Les organisations selon Simon, Nonaka et Takeuchi", *Bulletin de psychologie*, vol. 1, no. 475, pp. 149–156, 2005.

[DEM 18] DEMOULE J.-P., GARCIA D., SCHNAPP A., *Une histoire des civilisations*, La Découverte, Paris, 2018.

[DES 37] DESCARTES R., *Discours de la Méthode*, Imprimerie Ian Maire, 1637.

[DIA 92] DIAKOULAKI D., MAVROTAS G., PAPAYANNAKIS L., "A multicriteria approach for evaluating the performance of industrial firms", *Omega*, vol. 20, no. 4, pp. 467–474, 1992.

[DOR 96] DORIGO M., MANIEZZO V., COLORNI A., "Ant system: Optimization by a colony of cooperating agents", *IEEE Transactions on Systems, Man, and Cybernetics, Part B (Cybernetics)*, vol. 26, no. 1, pp. 29–41, 1996.

[DUB 83] DUBOIS D., PRADE H., "Unfair coins and necessity measures: Towards a possibilistic interpretation of histogram", *Fuzzy Sets and Systems*, no. 10, pp. 15–20, 1983.

[DUR 35] DURANT W., DURANT A., *The Story of Civilization*, 11 volumes, Simon and Schuster, New York, 1935.

[DYE 05] DYERS J.S., FIGUEIRA J., GRECO S. et al., *Multiple Criteria Decision Analysis. State of the Art Surveys*, Springer, Berlin, 2005.

[EAS 73] EASTON A., *Complex Managerial Decisions Involving Multiple Objectives*, John Wiley & Sons, New York, 1973.

[EDW 71] EDWARDS W., "Social utilities", *Engineering Economist, Summer Symposium Series*, no. 6, pp. 119–129, 1971.

[EHR 08] EHRGOTT M., "Multiobjective optimization", *AI Magazine*, 2008.

[FER 21] FERREIRA F.A.F., SANTOS S.P., "Two decades on the MACBETH approach: A bibliometric analysis", *Annals of Operations Research*, no. 296, pp. 901–925, 2021.

[FIG 05a] FIGUEIRA J.R., GRECO S., EHRGOTT S., *Multiple Criteria Decision Analysis. State of the Art Surveys*, Springer, Berlin, 2005.

[FIG 05b] FIGUEIRA J., MOUSSEAU V., ROY B., "ELECTRE methods", in FIGUEIRA J., GRECO S., EHRGOTT M. (eds), *Multiple Criteria Decision Analysis. State of the Art Surveys*, Springer, Berlin, 2005.

[FIS 67] FISHBURN P.C., "Methods of estimating additive utilities", *Management Science*, vol. 13, no. 7, pp. 435–453, 1967.

[FIS 70] FISHBURN P.C., *Utility Theory for Decision Making*, John Wiley & Sons, New York, 1970.

[FIS 81] FISHBURN P.C., "Subjective expected utility: A review of normative theories", *Theory and Decision*, no. 13, pp. 139–199, 1981.

[FRA 72] FRANKLIN B., From Benjamin Franklin to Joseph Priestley, Letter, Library of Congress, September 19, 1772.

[FRI 04] FRIEDBERG E., *La décision*, Éditions R&O Multimedia, Paris, 2004.

[GAG 01] GAGNÉ C., GRAVEL M., PRICE W.L., "Optimisation par colonie de fourmis pour un problème d'ordonnancement industriel avec temps de réglages dépendants de la séquence", *3e Conférence Francophone de MOdélisation et SIMulation*, Troyes, 2001.

[GAN 16] GANDY J.-M., PARIS F., *Établir mes documents ISO 9001 version 2015 : le couteau suisse de la Qualité*, AFNOR, Paris, 2016.

[GAR 11] GARDEUX V., Conception d'heuristiques d'optimisation pour les problèmes de grande dimension : application à l'analyse de données de puces à ADN, PhD thesis, Université Paris-Est, 2011.

[GEO 68] GEOFFRION A.M., "Proper efficiency and the theory of vector maximization", *Journal of Mathematical Analysis and Applications*, vol. 22, no. 3, pp. 618–630, 1968.

[GEO 71] GEORGESCU-ROEGEN N., *The Entropy Law and the Economic Process*, Harvard University Press, Cambridge, 1971.

[GER 97] GERNET J., "Le pouvoir d'État en Chine", *Actes de la recherche en sciences sociales*, no. 118, pp. 19–27, 1997.

[GIR 09] GIRAUD G., *La théorie des jeux*, Flammarion, Paris, 2009.

[GLO 86] GLOVER F., "Future paths for integer programming and links to artificial intelligence", *Computers & Operations Research*, vol. 13, no. 5, pp. 533–549, 1986.

[GOV 16] GOVINDAN K., JEPSEN M.B., "ELECTRE: A comprehensive literature review on methodologies and applications", *European Journal of Operational Research*, vol. 250, no. 1, pp. 1–29, 2016.

[GRA 05a] GRABISCH M., "Une approche constructive de la décision multicritère", *Traitement du Signal*, vol. 22, no. 4, pp. 321–337, 2005.

[GRA 05b] GRABISCH M., LABREUCHE C., "Fuzzy measures and integrals in MCDA", in FIGUEIRA J., GRECO S., EHRGOTT M. (eds), *Multiple Criteria Decision Analysis. State of the Art Surveys*, Springer, Berlin, 2005.

[GRA 15] GRAHAM A.C., *Disputers of the Tao: Philosophical Argument in Ancient China*, Open Court, Chicago, 2015.

[GRE 08] GRECO S., MOUSSEAU V., SLOWINSKI R., "Ordinal regression revisited: Multiple criteria ranking using a set of additive value functions", *European Journal of Operational Research*, vol. 191, no. 2, pp. 416–436, 2008.

[GRI 83] GRIFFITHS B., *Cosmic Revelation: The Hindu Way to God*, Templegate Publishers, Springfield, 1983.

[GUI 98] GUITOUNI A., MARTEL J.-M., "Tentative guidelines to help choosing an appropriate MCDA method", *European Journal of Operational Research*, vol. 109, no. 2, pp. 501–521, 1998.

[HAM 05] HAMDOUCH A., "Émergence et légitimité des institutions, coordination économique et nature de la rationalité des agents", *Innovation – The European Journal of Social Science Research*, vol. 18, no. 2, pp. 227–259, 2005.

[HAS 84] HASSE M., *La Grande Étude*, Éditions du Cerf, Paris, 1984.

[HER 92] HERBERT J., VARENNE J., *Vocabulaire de l'Hindouisme*, Dervy-Livres, Paris, 1992.

[HOL 75] HOLLAND J.H., *Adaptation in Natural and Artificial Systems*, University of Michigan Press, Ann Arbor, 1975.

[HWA 94] HWANG C.L., YOON K., *Multiple Attribute Decision Making: Methods and Applications.*, Springer-Verlag, New York, 1994.

[IMA 92] IMAI M., *Kaizen : la clé de la compétitivité japonaise*, Eyrolles, Paris, 1992.

[ISE 93] ISEN A.M., "Positive affect and decision making", in LEWIS M., HAVILAND J.M. (eds), *Handbook of Emotions*, Guilford Press, New York, 1993.

[ISH 11] ISHIZAKA A., LABIB A., "Review of the main developments in the analytic hierarchy process", *Expert Systems with Applications*, vol. 38, no. 11, pp. 14336–14345, 2011.

[ISO 01] ISO, Qualité et systèmes de management ISO 9000, version 2000, Standard, 2001.

[JAC 82] JACQUET-LAGREZE E., SISKOS J., "Assessing a set of additive utility functions for multicriteria decision-making, the UTA method", *European Journal of Operational Research*, vol. 10, no. 2, pp. 151–164, 1982.

[JAC 96] JACOT J.H., MICAELLI J.P., "La question de la performance globale", in JACOT J.H., MICAELLI J.P. (eds), *La performance économique en entreprise*, Hermès, Paris, 1996.

[JAV 89] JAVARY C., JING L.Y., *Le Grand Livre du Yin et du Yang*, Éditions du Cerf, Paris, 1989.

[JAY 18] JAY LEE J., DAVARI H., SINGH J. et al., "Industrial Artificial Intelligence for industry 4.0-based manufacturing systems", *Manufacturing Letters*, no. 18, pp. 20–23, 2018.

[KAH 79] KAHNEMAN A.D., TVERSKY A., "Prospect theory: An analysis of decision under risk", *Econometrica*, vol. 47, no. 2, pp. 263–291, 1979.

[KAR 13] KARANDEA P., CHAKRABORTY S., "Using MACBETH method for supplier selection in manufacturing environment", *International Journal of Industrial Engineering Computations*, no. 4, pp. 259–272, 2013.

[KEE 76] KEENEY R.L., RAIFFA H., *Decisions with Multiple Objectives: Preferences and Value Tradeoffs*, John Wiley & Sons, New York, 1976.

[KIR 83] KIRKPATRICK S., GELATT JR. C.D., VECCHI M.P., "Optimization by simulated annealing", *Science*, no. 220, pp. 671–680, 1983.

[KRA 71] KRANTZ D.H., LUCE R.D., SUPPES P. et al., *Foundations of Measurement, Vol. I. Additive and Polynomial Representations*, Academic Press, New York, 1971.

[LAO 15] LAO TSEU, *Tao-töking, livre sacré de la Voie et de la Vertu*, Gallimard, Paris, 2015.

[LAR 21] LAROUSSE, *Dictionnaire de français monolingue*, Larousse, Paris, 2021.

[LAU 10] LAURAS M., MARQUES G., GOURC D., "Towards a multi-dimensional project Performance Measurement System", *Decision Support Systems*, vol. 48, no. 2, pp. 342–353, 2010.

[LAZ 91] LAZARUS R.S., *Emotion and Adaptation*, Oxford University Press, Oxford, 1991.

[LEM 73] LE MOIGNE J.L., *Les systèmes d'information dans les organisations, vol. 1*, PUF, Paris, 1973.

[LEM 94] LE MOIGNE J.L., *La théorie du système général : théorie de la modélisation*, PUF, Paris, 1994.

[LET 21] LE TELLIER M., BERRAH L., CLIVILLÉ V. et al., "Using MACBETH for the performance expression of a mixed-use ecopark", *Journal of Multi-Criteria Decision Analysis*, no. 3, pp. 3–17, 2021.

[LOP 12] LOPEZ E., *Le grand livre de l'histoire des civilisations*, Eyrolles, Paris, 2012.

[MAR 58] MARCH J.G., SIMON H.A., *Organizations*, John Wiley & Sons, New York, 1958.

[MAR 78] MARCH J.G., "Bounded rationality, ambiguity, and the engineering of choice", *The Bell Journal of Economics*, vol. 9, no. 2, pp. 587–608, 1978.

[MAR 96] MARTIN H.J., *Histoire et pouvoirs de l'écrit*, Albin Michel, Paris, 1996.

[MAR 05] MARTEL J.-M., MATARAZZO B., "Other outranking approaches", in FIGUEIRA J., GRECO S., EHRGOTT M. (eds), *Multiple Criteria Decision Analysis. State of the Art Surveys*, Springer, Berlin, 2005.

[MAR 15] MARDANI A., JUSOH A., NOR K.M. et al., "Multiple criteria decision-making techniques and their applications – A review of the literature from 2000 to 2014", *Economic Research-Ekonomska Istrazivanja*, vol. 28, no. 1, pp. 516–571, 2015.

[MAS 96] MASPERO H., BALAZS E., *Histoire et institutions de la Chine ancienne. L'Antiquité. L'empire des Ts'in et des Han, Annales du Musée Guimet, Bibliothèque d'études*, volume LXXIII, PUF, Paris, 1996.

[MAS 00] MASCART J., *La vie et les travaux du chevalier de Borda, 1733–1799 : épisodes de la vie scientifique au XVIIIe siècle*, Presses de l'Université de Paris-Sorbonne, 2000.

[MAT 87] MATHIEU R., "Chamanes et chamanisme en Chine ancienne", *L'Homme*, no. 101, pp. 10–34, 1987.

[MAY 94] MAYSTRE L.Y., PICTET J., SIMOS J., *Méthodes multicritères ELECTRE. Description, conseils pratiques et cas d'application à la gestion environnementale*, Presses Polytechniques et Universitaire Romandes, Lausanne, 1994.

[MAY 10] MAYAG B., Élaboration d'une démarche constructive prenant en compte les interactions entre critères en aide multicritère à la décision, PhD thesis, Université de Paris 1, 2010.

[MER 16] MERKE A., *Le Principe de l'action humaine selon Démosthène et Aristote Hairesis – Prohairesis*, Les belles lettres, Paris, 2016.

[MIL 56] MILLER G., "The magical number seven, plus or minus two: Some limits on our capacity for processing information", *Psychological Review*, vol. 63, no. 2, pp. 343–352, 1956.

[MIN 59] MINSKY M., "Some methods of heuristic programming and artificial intelligence", *Proceedings of the Symposium on Mechanization of Thought Processes*, 1959.

[MIN 76] MINTZBERG H., RAISINGHANI D., THÉORÊT A., "The structure of 'unstructured' decision processes", *Administrative Science Quarterly*, vol. 21, no. 2, pp. 246–275, 1976.

[MOR 71] MORTON M.S., Management decision systems: Computer-based support for decision making, Document, Division of Research, Graduate School of Business Administration, Harvard University, Boston, 1971.

[MOR 17] MOREL P.M., "Vertu éthique et rationalité pratique chez Aristote. Note sur la notion d'hexis proairetikê", *Philonsorbonne*, no. 11, pp. 141–153, 2017.

[MOU 11] MOUINE M., Combinaison de deux méthodes d'analyse de sensibilité, Thesis, Faculté des études supérieures de l'Université Laval, Quebec, 2011.

[NAS 50a] NASH J.F., "The bargaining problem", *Econometrica*, no. 18, pp. 155–162, 1950.

[NAS 50b] NASH J.F., "Equilibrium points in n-person games", *Proceedings of the National Academy of Sciences of the United States of America*, 1950.

[NAS 51] NASH J., "Non-cooperative games", *Annals of Mathematics*, vol. 54, no. 2, pp. 286–295, 1951.

[NAU 09] NAU F., *La sagesse d'Ahiqar*, Letouzey et Ané, Paris, 1909.

[NEW 58] NEWELL A., SHAW J.C., SIMON H.A., "Elements of a theory of human problem solving", *Psychological Review*, vol. 65, no. 3, pp. 151–166, 1958.

[NIT 01] NITYABODHANANDA S., *Mythes et religions de l'Inde*, Maisonneuve & Larose, Paris, 2001.

[NON 91] NONAKA I., "The knowledge-creating company", *Harvard Business Review*, November–December, 1991.

[NON 95] NONAKA I., TAKEUCHI H., *The Knowledge-Creating Company: How Japanese Companies Create the Dynamics of Innovation*, Oxford University Press, 1995.

[NON 08] NONAKA I., TOYAMA R., HIRATA T., *Managing Flow: A Process Theory of the Knowledge-Based Firm*, Palgrave Macmillan, New York, 2008.

[NUR 10] NURMI H., "Voting systems for social choice", in KILGOUR D.M., EDEN C. (eds), *Handbook of Group Decision and Negotiation*, Springer, Berlin, 2010.

[OHN 88] OHNO T., *Toyota Production System: Beyond Large-Scale Production*, Productivity Press, New York 1988.

[PAR 09] PARETO V., *Manuel d'économie politique*, Hachette, Paris, 1909.

[PIN 86] PINFIELD L.T., "A field evaluation of perspectives on organizational decision making", *Administrative Science Quarterly*, vol. 31, no. 3, pp. 365–388, 1986.

[PIR 18] PIRENNE-DELFORGE V., *Le polythéisme grec comme objet d'histoire*, Fayard/Collège de France, Paris, 2018.

[PLA 97] PLATO, *Complete Works*, Hackett Publishing Company, Indianapolis, 1997.

[POM 87] POMIAN J., "Aux origines de l'intelligence artificielle : H.A. Simon en père fondateur", *Quaderni*, no. 1, pp. 9–25, 1987.

[POM 12] POMEROL J.-C., *Decision-Making and Action*, ISTE Ltd, London, and John Wiley & Sons, Hoboken, 2012.

[PRA 64] PRATT J.W., "Risk aversion in the small and in the large", *Econometrica*, vol. 32, no. 1/2, pp. 122–136, 1964.

[RAM 31] RAMSEY F.P., "Truth and probability", in BRAITHWAITE R.B. (ed.), *The Foundations of Mathematics and other Logical Essays*, Kegan, Paul, Trench, Trubner & Co., London, 1931.

[REN 78] RENOU L., MALAMOUD C., *L'Inde fondamentale*, Hermann, Paris, 1978.

[ROA 21] ROADEF, Le livre banc de la recherche opérationnelle en France, White book, ROADEF, Paris, 2021.

[ROB 91a] ROBIN C., "Cités, royaumes et empires de l'Arabie avant l'Islam", *Revue des mondes musulmans et de la Méditerranée*, vol. 61, pp. 45–54, 1991.

[ROB 91b] ROBINET I., *Histoire du Taoïsme des origines au XIVe siècle*, Éditions du Cerf, Paris, 1991.

[ROM 01] ROMELAER P., LAMBERT G., "Décisions d'investissement et rationalités", in CHARREAUX G. (ed.), *Images de l'investissement. Au-delà de l'évaluation financière : une lecture organisationnelle et stratégique*, Vuibert, Paris, 2001.

[ROY 68a] ROY B., "Classement et choix en présence de points de vue multiples (la méthode ELECTRE)", *La Revue d'Informatique et de Recherche Opérationnelle (RIRO)*, no. 8, pp. 57–75, 1968.

[ROY 68b] ROY B., "Il faut désoptimiser la recherche opérationnelle", *Bulletin de l'AFIR*, no. 7, 1968.

[ROY 85] ROY B., *Méthodologie multicritères d'aide à la décision*, Economica, Paris, 1985.

[ROY 91] ROY B., "The outranking approach and the foundations of ELECTRE methods", *Theory and Decision*, no. 31, pp. 49–73, 1991.

[ROY 93] ROY B., BOUYSSOU D., *Aide Multicritère à la Décision : méthodes et cas*, Economica, Paris, 1993.

[ROY 97] ROY B., VANDERPOOTEN D., "An overview on 'The European school of MCDA': Emergence, basic features and current works", *European Journal of Operational Research*, vol. 99, no. 1, pp. 26–27, 1997.

[ROY 00] ROY B., "L'aide à la décision aujourd'hui : que devrait-on en attendre ?", in DAVID A., HATCHUEL A., LAUFER R. (eds), *Les nouvelles fondations des Sciences de gestion – Éléments d'épistémologie de la recherche en management*, Vuibert, Paris, 2000.

[ROY 05] ROY B., "Paradigms and Challenges", in FIGUEIRA J., GRECO S., EHRGOTT M. (eds), *Multiple Criteria Decision Analysis. State of the Art Surveys*, Springer, Berlin, 2005.

[SAA 77] SAATY T., "A scaling method for priorities in hierarchical structures", *Journal of Mathematical Psychology*, no. 15, pp. 234–281, 1977.

[SAA 84] SAATY T., *Décider face à la complexité : une approche analytique multi-critère d'aide à la décision*, Entreprise moderne d'édition, Paris, 1984.

[SAA 05] SAATY T., "The analytical hierarchy and analytic network process for the measurement of intangible criteria and for decision-making", in FIGUEIRA J., GRECO S., EHRGOTT M. (eds), *Multiple Criteria Decision Analysis. State of the Art Surveys*, Springer, Berlin, 2005.

[SAV 54] SAVAGE L.J., *The Foundations of Statistics*, John Wiley & Sons, New York, 1954.

[SCH 85] SCHÄRLIG A., *Décider sur plusieurs critères : panorama de l'aide à la décision multicritère*, Presses Polytechniques et Universitaire Romandes, Lausanne, 1985.

[SCH 94] SCHENKERMAN S., "Avoiding rank reversal in AHP decision-support models", *European Journal of Operational Research*, vol. 74, no. 3, pp. 407–419, 1994.

[SCH 96] SCHÄRLIG A., "The case of the vanishing optimum", *Journal of Multi-Criteria Decision Analysis*, vol. 5, no. 2, pp. 160–164, 1996.

[SCH 03] SCHUMANN-ANTELME R., *Dictionnaire illustré des dieux de l'Égypte*, Éditions du Rocher, Monaco, 2003.

[SHA 12] SHAH L., Value/risk based performance evaluation of industrial systems, PhD thesis, Arts et Métiers ParisTech – Centre de Metz, 2012.

[SIM 47] SIMON H.A., *Administrative Behavior: A Study of Decision-Making Processes in Administrative Organization*, The Free Press, New York, 1947.

[SIM 55] SIMON H.A., "A behavioral model of rational choice source", *The Quarterly Journal of Economics*, vol. 69, no. 1, pp. 99–118, 1955.

[SIM 60] SIMON H.A., *The New Science of Management Decision*, Harper & Brothers, New York, 1960.

[SIM 76] SIMON H.A., *From Substantive to Procedural Rationality*, Cambridge University Press, Cambridge, 1976.

[SIP 10] SIPAHI S., TIMOR M., "The analytic hierarchy process and analytic network process: An overview of applications", *Management Decision*, vol. 48, no. 5, pp. 775–808, 2010.

[SMI 76] SMITH A., *An Inquiry into the Nature and Causes of the Wealth of Nations*, Methuen and Co, London, 1776.

[STE 46] STEVENS S.S., "On the theory of scales of measurement", *Science*, no. 103, pp. 677–680, 1946.

[TAK 86] TAKEUCHI H., NONAKA I., "The new new product development game – Stop running the relay race and take up rugby", *Harvard Business Review*, pp. 137–146, January–February 1986.

[TAT 64] TATON R., *Histoire générale des sciences. Tome III : La science contemporaine, vol. II. Le XXe siècle*, PUF, Paris, 1964.

[TEI 19] TEIXEIRA DE ALMEIDA A., COSTA MORAIS D., NURMI H., *Systems, Procedures and Voting Rules in Context: A Primer for Voting Rule Selection*, Springer, Berlin, 2019.

[THE 20] THEBAULT M., CLIVILLÉ V., BERRAH L. et al., "Multicriteria roof sorting for the integration of photovoltaic systems in urban environments", *Sustainable Cities and Society*, vol. 60, no. 102259, 2020.

[THU 17] THURSBY G., MITTAL S., *Religions of India: An Introduction*, Routledge, London, 2017.

[TIC 76] TICHKIEWITCH S., "Capitalization and reuse of forging knowledge in integrated design", in BERNARD A. (ed.), *Method and Tools for the Effective Knowledge Management in Product Life Cycle*, Springer, Berlin, 1976.

[TRI 00] TRIANTAPHYLLOU E., *Multi-Criteria Decision Making Methods: A Comparative Study*, Springer, Berlin, 2000.

[TSO 06] TSOUKIAS A., "De la théorie de la décision à l'aide à la décision", in BOUYSSOU D., DUBOIS D., PIRLOT M. et al. (eds), *Concepts et méthodes pour l'aide à la décision, volume 2 : Outils de modélisation*, Hermès-Lavoisier, Paris, 2006.

[VAN 86] VANSNICK J.-C., "On the problem of weights in multiple criteria decision making (the noncompensatory approach)", *European Journal of Operational Research*, vol. 24, no. 2, pp. 288–294, 1986.

[VAN 08] VAN HOOREBEKE D., "L'émotion et la prise de décision", *Revue française de gestion*, vol. 2, no. 182, pp. 33–44, 2008.

[VAR 97] VARGAS L.G., "Why the multiplicative AHP is invalid: A practical example", *Journal of Multi-Criteria Decision Analysis*, no. 6, pp. 169–170, 1997.

[VER 03] VERDIER F., *Passagère du silence. Dix ans d'initiation en Chine*, Albin Michel, Paris, 2003.

[VIN 92] VINCKE P., *Multicriteria Decision-Aid*, John Wiley & Sons, Chichester, 1992.

[VOG 95] VOGT H., *Leçons sur la résolution algébrique des équations*, Cornell University Library, Ithaca, 1895.

[VON 44] VON NEUMANN J., MORGENSTERN O., *Theory of Games and Economic Behavior*, Princeton University Press, 1944.

[YU 92] YU W., Aide multicritère à la décision dans le cadre de la problématique du tri : concepts, méthodes et applications, PhD thesis, Université Paris 9, 1992.

[ZAD 78] ZADEH L., "Fuzzy sets as a basis for a theory of possibility", *Fuzzy Sets and Systems*, no. 1, pp. 3–28, 1978.

[ZEL 73] ZELENY M., COCHRANE J.L., "A priori and a posteriori goals in macro-economic policy making", in ZELENY M., COCHRANE J.L. (eds), *Multiple Criteria Decision Making*, University of South Carolina Press, Columbia, 1973.

Index

A, B

action (*see also* alternative), 2, 9, 14, 18, 32, 34, 37, 41, 57–59, 61, 63, 67, 96, 98, 108, 122
  plan, 7, 57–61, 124, 126
aggregation (*see also* American school), 96, 98, 127, 128, 146–150, 152, 153, 164, 168–171, 178, 184
  operator, 149, 150, 169, 171, 186
AHP (analytic hierarchy process), 149, 152–155, 157, 159–161, 165–169, 171–173, 175, 178, 184, 186
alternative (*see also* action, solution), 90, 91, 99–102, 104–110, 112–118, 121–123, 128, 129, 131, 133, 135, 136, 139, 143, 146, 150, 151, 153, 164, 166, 168, 169, 177, 178, 180, 182, 186, 187
artificial intelligence (AI), 83
attractiveness, 169–176, 178, 180–183, 187
attribute (*see also* criterion/criteria), 101–104, 106, 147
belief, 26, 41, 148
business, 43, 49, 53, 55, 81, 82, 94, 100, 110

C, D

choice, 26, 41, 42, 47, 49, 52, 53, 62–64, 66, 68, 74, 75, 79, 82, 85, 87, 91–95, 97, 99, 103, 104, 107, 109, 110, 122, 124, 126, 127, 130, 131, 134, 135, 146, 149, 151, 152, 165, 187, 188
choosing, 33, 41, 52, 66, 67, 82, 89, 91, 93–95, 99, 106, 107, 121, 126, 128, 187
civilization, 3, 5, 6, 8, 10, 12, 13, 16, 21, 25, 28, 34
classification (*see also* sorting), 67, 82, 97, 130
compare/comparison, 28, 38, 50, 62–64, 68, 72, 84, 90, 93, 94, 106, 108, 111, 113, 126, 127, 129–131, 137, 141, 143, 147, 148, 150, 152, 158, 162, 169, 171, 173, 178
compromise, 34, 40, 49, 51, 52, 59, 62, 65, 106, 186
concordance, 129–132, 135, 136, 138–140, 142, 144, 146
consensus, 5, 27, 28, 82
context (*see also* environment), 1, 2, 4, 5, 13, 21, 29, 33, 34, 42, 45, 47–50, 53, 60, 61, 67, 87, 89, 96, 131

credibility, 144, 146
criterion/criteria (*see also* attribute, MCDA), 49–54, 57, 58, 61, 67, 81, 93, 94, 101, 104–107, 113–117, 119, 122, 135–143, 147, 152, 154, 155, 157, 161, 162, 164, 168, 169, 171, 175, 178–180, 183, 186
   hierarchy of, 155
   multi-, 91, 94, 152
   weight of, 64, 139, 145, 148, 153, 161–164, 166, 167, 170, 171, 178–180, 182–186
debate (*see also* deliberation), 42, 66, 167
Decision Deck, 134, 147, 167
decision-aiding, 90, 95, 97, 98, 108, 126, 134, 146, 147, 166, 167, 186
deliberation (*see also* debate), 25, 39, 41, 65, 67
description (*see also* scoring), 4–6, 28, 62, 66, 71, 76, 88, 101, 168
design, 68–70, 73, 76–81, 85
diagnosis, 68–70, 74, 76, 84, 85
discordance, 129–132, 134–136, 138, 139, 141–144, 146

E, F

ELECTRE (*ELimination Et Choix Traduisant la REalité* – Elimination and Choice Translating Reality), 130, 131, 134–138, 140, 142–147, 153, 188
emotion, 15, 44, 71
environment (*see also* context), 2, 7, 10, 13, 17, 22, 39, 124
evaluation, 1, 11, 18, 71–73, 85, 87, 92, 149, 155
fear, 40, 41
formalization, 8, 12, 34, 42, 47, 71, 148

H, I

Hasse diagram (*see also* order relationship), 133, 145, 151, 165, 185
hesitation, 40, 42, 91
heuristics, 49, 51, 66, 81
history, 2, 5, 6, 28, 103
humankind, 5, 8, 15, 25, 29, 33, 34, 42, 88, 155
improvement, 21, 42, 47, 63, 69, 76, 121
incomparability, 80, 95, 126, 128–131, 148
inconsistency coefficient, 160
indifference, 99, 105, 112–114, 119, 129–132, 134, 137, 138, 146, 150, 172
intention, 1, 9, 17, 37, 38, 41, 47, 55, 63, 65, 71, 91, 100, 104, 111, 170, 187
intuition, 31, 42, 44, 74, 143, 186

K, L

knowledge, 18, 66, 70, 77, 78, 81, 152
linear program, 55, 175
logic, 3, 14, 42, 43, 58, 103

M, N

MACBETH (Measuring Attractiveness by a Categorical Based Evaluation TecHnique), 149, 167–177, 181–188
majority, 131, 144, 147
manager, 30, 67, 69–71, 74, 76, 80, 81, 94, 95, 99–103, 106, 113–117, 122, 128, 131–133, 136, 138, 141, 142, 145, 146, 150, 151, 154, 156–158, 161, 162, 164–166, 172, 173, 176, 177, 182–184

MCDA (multi-criteria decision-aiding), 92, 95–99, 101, 103, 104, 107, 109, 110, 112, 114, 120, 124–127, 130, 136, 146, 149, 152, 157, 166, 188
nature, 3, 7, 14, 17–19, 23, 27, 29, 33, 34, 39, 44, 45, 47–49, 61, 66, 88, 106, 110

O, P

objective, 1, 7, 37, 38, 40, 43, 45–50, 58, 60–68, 73, 75, 76, 110–112
operational research, 47, 48, 59
optimization, 40, 47, 49–51, 54, 55, 57, 58, 61, 65, 67, 72, 81, 97, 168
order, 17, 19, 92, 95, 106, 108, 111, 112, 114–118, 120, 121, 124, 126, 144, 146, 148, 150–153, 165, 166, 168, 185–187
   relationship (*see also* Hasse diagram), 108, 111, 112, 114, 174
outranking (*see also* European school), 97, 127, 129–139, 144, 148, 149
performance, 11, 13, 17, 22, 27, 34, 38, 57–59, 69, 74, 76, 79, 85, 101, 103, 135, 168
philosophy, 3, 13, 15, 20, 28, 62, 92, 168
politics, 24, 71, 94, 96
pre-order, 120, 121, 128, 131, 145, 150
preference, 62, 90, 92, 93, 104, 105, 108, 112–115, 117, 119, 126, 127, 129, 132, 137, 138, 140, 142, 143, 146, 147, 150, 152, 168, 172
priority, 108, 153, 155, 159, 164, 168, 184
process, 1, 9, 25, 32–34, 37–46, 56–61, 63, 65, 67, 68, 74, 75, 77–80, 83–91, 96, 98, 100, 107, 112, 114, 124–126, 188

Q, R

quantification, 46, 104, 110, 130, 135, 136, 147, 148, 153, 166
rank/ranking, 66, 106–109, 111, 114, 115, 121, 130, 131, 134, 146, 150–152, 168, 186
rationality, 9, 39, 41, 44–47, 51, 53, 56, 57, 60–67, 80, 83–87, 89, 91, 92, 95, 98, 114, 166, 169
reason, 3, 24, 32, 39, 44, 65
reflection, 3, 11, 18, 27, 28, 41, 45, 63, 67, 70, 89
relative magnitude, 153, 155
responsibility, 33
risk, 7, 54, 55, 61, 159

S, T

satisfaction level, 107, 141
scale, 61, 136, 141, 169, 177, 182–184
school
   American (*see also* aggregation), 98, 109, 128
   European (*see also* outranking), 98, 127, 128
science/scientific, 9, 21, 24, 25, 28, 32, 39, 48, 60, 66, 83, 152
score, 73, 109, 111, 150, 168, 169, 171, 172, 175, 176, 178, 179, 182, 183, 185, 186
scoring (*see also* description), 107, 109–111
society, 3, 15, 20
solution (*see also* alternative), 3, 33, 42, 44–46, 49–55, 57, 58, 60–74, 76, 78–83, 85–89, 91, 94, 95, 98, 99, 101, 108, 134, 147, 149, 152
sorting (*see also* classification), 107, 109–111, 134
state, 67, 69, 71, 72, 112, 114

system, 1–3, 7–12, 14–17, 19,
  21–23, 26–28, 38, 40, 46, 49–51,
  55, 60, 68–72, 76, 82, 88, 94, 100,
  109, 133, 151
systemic, 88
territory, 24
thought, 6, 10, 41, 60, 84, 88, 170
threshold, 71, 138, 140–143

U, V, W

uncertainty, 40, 53, 61, 148
UTA (additive utilities), 149, 188
utility, 33, 96, 98, 147, 149, 168
variable, 46, 50, 54, 55, 93, 101, 104
veto, 134, 140–143, 146
voting, 96, 97
WAM (weighted arithmetic mean),
  148, 153, 164, 169–171, 184, 186

Other titles from

in

Systems and Industrial Engineering – Robotics

2022

AMARA Yacine, BEN AHMED Hamid, GABSI Mohamed
Hybrid Excited Synchronous Machines: Topologies, Design and Analysis

BOURRIÈRES Jean-Paul, PINÈDE Nathalie, TRAORÉ Mamadou Kaba, ZACHAREWICZ Grégory
From Logistic Networks to Social Networks: Similarities, Specificities, Modeling, Evaluation

DEMOLY Frédéric, ANDRÉ Jean-Claude
4D Printing 1: Between Disruptive Research and Industrial Applications
4D Printing 2: Between Science and Technology

HAJJI Rafika, JARAR OULIDI Hassane
Building Information Modeling for a Smart and Sustainable Urban Space

KROB Daniel
Model-based Systems Architecting: Using CESAM to Architect Complex Systems (Volume 3 - Systemes of Systems Complexity Set)

LOUIS Gilles
Dynamics of Aircraft Flight

2020

BRON Jean-Yves
System Requirements Engineering

KRYSINSKI Tomasz, MALBURET François
Energy and Motorization in the Automotive and Aeronautics Industries

PRINTZ Jacques
System Architecture and Complexity: Contribution of Systems of Systems to Systems Thinking

2019

ANDRÉ Jean-Claude
Industry 4.0: Paradoxes and Conflicts

BENSALAH Mounir, ELOUADI Abdelmajid, MHARZI Hassan
Railway Information Modeling RIM: The Track to Rail Modernization

BLUA Philippe, YALAOU Farouk, AMODEO Lionel, DE BLOCK Michaël, LAPLANCHE David
Hospital Logistics and e-Management: Digital Transition and Revolution

BRIFFAUT Jean-Pierre
From Complexity in the Natural Sciences to Complexity in Operations Management Systems
(Systems of Systems Complexity Set – Volume 1)

BUDINGER Marc, HAZYUK Ion, COÏC Clément
Multi-Physics Modeling of Technological Systems

FLAUS Jean-Marie
Cybersecurity of Industrial Systems

JAULIN Luc
Mobile Robotics – Second Edition Revised and Updated

KUMAR Kaushik, DAVIM Paulo J.
Optimization for Engineering Problems

TRIGEASSOU Jean-Claude, MAAMRI Nezha
Analysis, Modeling and Stability of Fractional Order Differential Systems 1: The Infinite State Approach
Analysis, Modeling and Stability of Fractional Order Differential Systems 2: The Infinite State Approach

VANDERHAEGEN Frédéric, MAAOUI Choubeila, SALLAK Mohamed, BERDJAG Denis
Automation Challenges of Socio-technical Systems

2018

BERRAH Lamia, CLIVILLÉ Vincent, FOULLOY Laurent
Industrial Objectives and Industrial Performance: Concepts and Fuzzy Handling

GONZALEZ-FELIU Jesus
Sustainable Urban Logistics: Planning and Evaluation

GROUS Ammar
Applied Mechanical Design

LEROY Alain
Production Availability and Reliability: Use in the Oil and Gas Industry

MARÉ Jean-Charles
Aerospace Actuators 3: European Commercial Aircraft and Tiltrotor Aircraft

MAXA Jean-Aimé, BEN MAHMOUD Mohamed Slim, LARRIEU Nicolas
Model-driven Development for Embedded Software: Application to Communications for Drone Swarm

MBIHI Jean
Analog Automation and Digital Feedback Control Techniques
Advanced Techniques and Technology of Computer-Aided Feedback Control

MORANA Joëlle
Logistics

SIMON Christophe, WEBER Philippe, SALLAK Mohamed
Data Uncertainty and Important Measures
(Systems Dependability Assessment Set – Volume 3)

TANIGUCHI Eiichi, THOMPSON Russell G.
City Logistics 1: New Opportunities and Challenges
City Logistics 2: Modeling and Planning Initiatives
City Logistics 3: Towards Sustainable and Liveable Cities

ZELM Martin, JAEKEL Frank-Walter, DOUMEINGTS Guy, WOLLSCHLAEGER Martin
Enterprise Interoperability: Smart Services and Business Impact of Enterprise Interoperability

2017

ANDRÉ Jean-Claude
From Additive Manufacturing to 3D/4D Printing 1: From Concepts to Achievements
From Additive Manufacturing to 3D/4D Printing 2: Current Techniques, Improvements and their Limitations
From Additive Manufacturing to 3D/4D Printing 3: Breakthrough Innovations: Programmable Material, 4D Printing and Bio-printing

ARCHIMÈDE Bernard, VALLESPIR Bruno
Enterprise Interoperability: INTEROP-PGSO Vision

CAMMAN Christelle, FIORE Claude, LIVOLSI Laurent, QUERRO Pascal
Supply Chain Management and Business Performance: The VASC Model

FEYEL Philippe
Robust Control, Optimization with Metaheuristics

MARÉ Jean-Charles
Aerospace Actuators 2: Signal-by-Wire and Power-by-Wire

POPESCU Dumitru, AMIRA Gharbi, STEFANOIU Dan, BORNE Pierre
Process Control Design for Industrial Applications

RÉVEILLAC Jean-Michel
Modeling and Simulation of Logistics Flows 1: Theory and Fundamentals
Modeling and Simulation of Logistics Flows 2: Dashboards, Traffic Planning and Management
Modeling and Simulation of Logistics Flows 3: Discrete and Continuous Flows in 2D/3D

2016

ANDRÉ Michel, SAMARAS Zissis
Energy and Environment
(Research for Innovative Transports Set - Volume 1)

AUBRY Jean-François, BRINZEI Nicolae, MAZOUNI Mohammed-Habib
Systems Dependability Assessment: Benefits of Petri Net Models (Systems Dependability Assessment Set - Volume 1)

BLANQUART Corinne, CLAUSEN Uwe, JACOB Bernard
Towards Innovative Freight and Logistics (Research for Innovative Transports Set - Volume 2)

COHEN Simon, YANNIS George
Traffic Management (Research for Innovative Transports Set - Volume 3)

MARÉ Jean-Charles
Aerospace Actuators 1: Needs, Reliability and Hydraulic Power Solutions

REZG Nidhal, HAJEJ Zied, BOSCHIAN-CAMPANER Valerio
Production and Maintenance Optimization Problems: Logistic Constraints and Leasing Warranty Services

TORRENTI Jean-Michel, LA TORRE Francesca
Materials and Infrastructures 1 (Research for Innovative Transports Set - Volume 5A)
Materials and Infrastructures 2 (Research for Innovative Transports Set - Volume 5B)

WEBER Philippe, SIMON Christophe
Benefits of Bayesian Network Models
(Systems Dependability Assessment Set – Volume 2)

YANNIS George, COHEN Simon
Traffic Safety (Research for Innovative Transports Set - Volume 4)

2015

AUBRY Jean-François, BRINZEI Nicolae
Systems Dependability Assessment: Modeling with Graphs and Finite State Automata

BOULANGER Jean-Louis
CENELEC 50128 and IEC 62279 Standards

BRIFFAUT Jean-Pierre
E-Enabled Operations Management

MISSIKOFF Michele, CANDUCCI Massimo, MAIDEN Neil
Enterprise Innovation

2014

CHETTO Maryline
Real-time Systems Scheduling
Volume 1 – Fundamentals
Volume 2 – Focuses

DAVIM J. Paulo
Machinability of Advanced Materials

ESTAMPE Dominique
Supply Chain Performance and Evaluation Models

FAVRE Bernard
Introduction to Sustainable Transports

GAUTHIER Michaël, ANDREFF Nicolas, DOMBRE Etienne
Intracorporeal Robotics: From Milliscale to Nanoscale

MICOUIN Patrice
Model Based Systems Engineering: Fundamentals and Methods

MILLOT Patrick
Designing Human–Machine Cooperation Systems

NI Zhenjiang, PACORET Céline, BENOSMAN Ryad, RÉGNIER Stéphane
Haptic Feedback Teleoperation of Optical Tweezers

OUSTALOUP Alain
Diversity and Non-integer Differentiation for System Dynamics

REZG Nidhal, DELLAGI Sofien, KHATAD Abdelhakim
Joint Optimization of Maintenance and Production Policies

STEFANOIU Dan, BORNE Pierre, POPESCU Dumitru, FILIP Florin Gh., EL KAMEL Abdelkader
Optimization in Engineering Sciences: Metaheuristics, Stochastic Methods and Decision Support

2013

ALAZARD Daniel
Reverse Engineering in Control Design

ARIOUI Hichem, NEHAOUA Lamri
Driving Simulation

CHADLI Mohammed, COPPIER Hervé
Command-control for Real-time Systems

DAAFOUZ Jamal, TARBOURIECH Sophie, SIGALOTTI Mario
Hybrid Systems with Constraints

FEYEL Philippe
Loop-shaping Robust Control

FLAUS Jean-Marie
Risk Analysis: Socio-technical and Industrial Systems

FRIBOURG Laurent, SOULAT Romain
Control of Switching Systems by Invariance Analysis: Application to Power Electronics

GROSSARD Mathieu, RÉGNIER Stéphane, CHAILLET Nicolas
Flexible Robotics: Applications to Multiscale Manipulations

GRUNN Emmanuel, PHAM Anh Tuan
Modeling of Complex Systems: Application to Aeronautical Dynamics

HABIB Maki K., DAVIM J. Paulo
Interdisciplinary Mechatronics: Engineering Science and Research Development

HAMMADI Slim, KSOURI Mekki
Multimodal Transport Systems

JARBOUI Bassem, SIARRY Patrick, TEGHEM Jacques
Metaheuristics for Production Scheduling

KIRILLOV Oleg N., PELINOVSKY Dmitry E.
Nonlinear Physical Systems

LE Vu Tuan Hieu, STOICA Cristina, ALAMO Teodoro, CAMACHO Eduardo F., DUMUR Didier
Zonotopes: From Guaranteed State-estimation to Control

MACHADO Carolina, DAVIM J. Paulo
Management and Engineering Innovation

MORANA Joëlle
Sustainable Supply Chain Management

SANDOU Guillaume
Metaheuristic Optimization for the Design of Automatic Control Laws

STOICAN Florin, OLARU Sorin
Set-theoretic Fault Detection in Multisensor Systems

2012

AÏT-KADI Daoud, CHOUINARD Marc, MARCOTTE Suzanne, RIOPEL Diane
Sustainable Reverse Logistics Network: Engineering and Management

BORNE Pierre, POPESCU Dumitru, FILIP Florin G., STEFANOIU Dan
Optimization in Engineering Sciences: Exact Methods

CHADLI Mohammed, BORNE Pierre
Multiple Models Approach in Automation: Takagi-Sugeno Fuzzy Systems

DAVIM J. Paulo
Lasers in Manufacturing

DECLERCK Philippe
Discrete Event Systems in Dioid Algebra and Conventional Algebra

DOUMIATI Moustapha, CHARARA Ali, VICTORINO Alessandro, LECHNER Daniel
Vehicle Dynamics Estimation using Kalman Filtering: Experimental Validation

GUERRERO José A, LOZANO Rogelio
Flight Formation Control

HAMMADI Slim, KSOURI Mekki
Advanced Mobility and Transport Engineering

MAILLARD Pierre
Competitive Quality Strategies

MATTA Nada, VANDENBOOMGAERDE Yves, ARLAT Jean
Supervision and Safety of Complex Systems

POLER Raul *et al.*
Intelligent Non-hierarchical Manufacturing Networks

TROCCAZ Jocelyne
Medical Robotics

YALAOUI Alice, CHEHADE Hicham, YALAOUI Farouk, AMODEO Lionel
Optimization of Logistics

ZELM Martin *et al.*
Enterprise Interoperability –I-EASA12 Proceedings

2011

CANTOT Pascal, LUZEAUX Dominique
Simulation and Modeling of Systems of Systems

DAVIM J. Paulo
Mechatronics

DAVIM J. Paulo
Wood Machining

GROUS Ammar
Applied Metrology for Manufacturing Engineering

KOLSKI Christophe
Human–Computer Interactions in Transport

LUZEAUX Dominique, RUAULT Jean-René, WIPPLER Jean-Luc
Complex Systems and Systems of Systems Engineering

ZELM Martin, *et al.*
Enterprise Interoperability: IWEI2011 Proceedings

2010

BOTTA-GENOULAZ Valérie, CAMPAGNE Jean-Pierre, LLERENA Daniel, PELLEGRIN Claude
Supply Chain Performance / Collaboration, Alignement and Coordination

BOURLÈS Henri, GODFREY K.C. Kwan
Linear Systems

BOURRIÈRES Jean-Paul
Proceedings of CEISIE'09

CHAILLET Nicolas, REGNIER Stéphane
Microrobotics for Micromanipulation

DAVIM J. Paulo
Sustainable Manufacturing

GIORDANO Max, MATHIEU Luc, VILLENEUVE François
Product Life-Cycle Management / Geometric Variations

LOZANO Rogelio
Unmanned Aerial Vehicles / Embedded Control

LUZEAUX Dominique, RUAULT Jean-René
Systems of Systems

VILLENEUVE François, MATHIEU Luc
Geometric Tolerancing of Products

2009

DIAZ Michel
Petri Nets / Fundamental Models, Verification and Applications

OZEL Tugrul, DAVIM J. Paulo
Intelligent Machining

PITRAT Jacques
Artificial Beings

2008

ARTIGUES Christian, DEMASSEY Sophie, NÉRON Emmanuel
Resources–Constrained Project Scheduling

BILLAUT Jean-Charles, MOUKRIM Aziz, SANLAVILLE Eric
Flexibility and Robustness in Scheduling

DOCHAIN Denis
Bioprocess Control

LOPEZ Pierre, ROUBELLAT François
Production Scheduling

THIERRY Caroline, THOMAS André, BEL Gérard
Supply Chain Simulation and Management

2007

DE LARMINAT Philippe
Analysis and Control of Linear Systems

DOMBRE Etienne, KHALIL Wisama
Robot Manipulators

LAMNABHI Françoise *et al.*
Taming Heterogeneity and Complexity of Embedded Control

LIMNIOS Nikolaos
Fault Trees

2006

FRENCH COLLEGE OF METROLOGY
Metrology in Industry

NAJIM Kaddour
Control of Continuous Linear Systems

Printed by BoD™in Norderstedt, Germany